MAINTENANCE PROGRAMMING

Improved Productivity Through Motivation

J. DANIEL COUGER

Distinguished Professor
Computer and Management Science
University of Colorado

MEL A. COLTER

Associate Professor
Management Science
University of Colorado

Prentice-Hall, Inc.

Englewood Cliffs, New Jersey 07632

Library of Congress Catalog Card Number: 84-62983

ISBN: 0-13-545450-6

Printed in the United States of America

10 9 8 7 6 5 4 3 2 1

ISBN 0-13-545450-6 01

Prentice-Hall International, Inc., *London*
Prentice-Hall of Australia Pty. Limited, *Sydney*
Editora Prentice-Hall do Brasil, Ltda., *Rio de Janeiro*
Prentice-Hall Canada Inc., *Toronto*
Prentice-Hall Hispanoamericana, S.A., *Mexico*
Prentice-Hall of India Private Limited, *New Delhi*
Prentice-Hall of Japan, Inc., *Tokyo*
Prentice-Hall of Southeast Asia Pte. Ltd., *Singapore*
Whitehall Books Limited, *Wellington, New Zealand*

TABLE OF CONTENTS

PREFACE

This report contains results of in-depth research on approaches to motivate programmers and analysts assigned to maintenance activities. In phase I of the reasearch, over 500 persons in 10 organizations completed the Couger-Zawacki (C-Z) diagnostic survey questionnaire for computer personnel. In phase II, on-site interviews were conducted with 104 persons (61 analysts and programmers and 43 supervisors/managers of system departments). Data were analyzed utilizing rigorous statistical methodology. The results indicate that productivity can be improved by motivation enhancement procedures.

Section 1 provides the background on the maintenance problem. Not only is more than 50% of the labor budget devoted to maintenance in the typical organization, but this work also is considered the least desirable by the large majority of personnel. That problem is elaborated in Chapter 1. Chapter 2 covers the general model of motivation and its application to computer personnel. The C-Z national data base of 6,000 computer personnel enabled establishing motivational norms for 18 different jobs within the field. Chapter 3 covers the research methodology for this project, which consisted of two instruments: 1) the C-Z diagnostic survey (the same one used to establish the national data base), and 2) the structured interview questionnaire on motivation.

Chapter 4 provides a management-level summary of the results of the research. Figure 4.1, repeated here, is perhaps the most "telling" result in the survey. As shown by the curve on the left, fix-it activities are especially de-motivating. However, as shown by the curve on the right side, every category of maintenance is de-motivating. This chart illustrates the huge potential for improvement. It shows that maintenance work ranges from 1/2 to 2/3 the motivating potential of other programming/analysis work. Since that situation applies to over half the work being done in the typical organization, the potential for improvement is staggering.

Fortunately, the research proved that maintenance jobs can be enhanced to the point that motivation levels are comparable to those in new system development. Such a condition already existed in some of the firms selected for the research and we were able to document the differences in those jobs. In addition, we have been involved in other job enhancement projects that raised the motivation level of maintenance to that of new development. Chapter 4 summarizes these results and provides recommendations on how to enrich maintenance work.

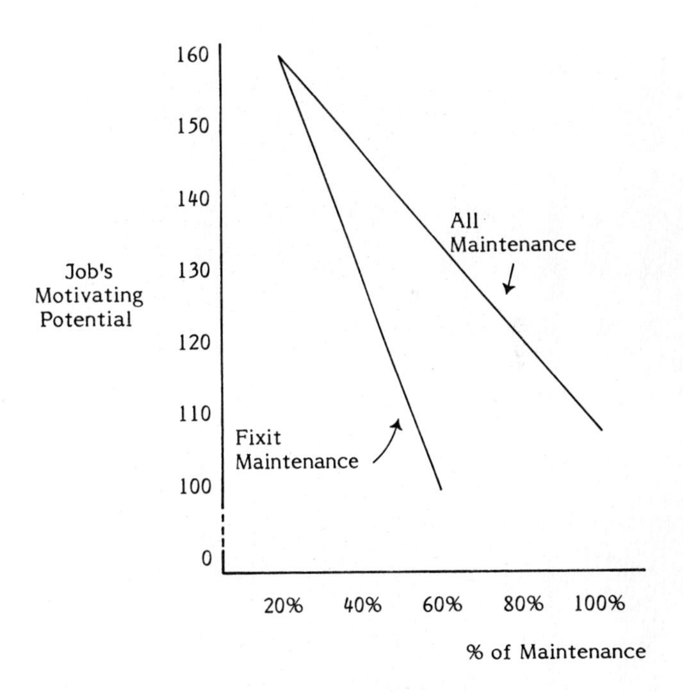

Effect of Percent of Maintenance On Job's Motivating Potential

Chapter 5 provides 14 actual cases illustrating application of motivation principles to varying maintenance situations.

Chapters 6 and 7 provide the aggregate results of the research. The summary data on the 10 organizations is compared to national norms and the interviewed professionals are compared to both the national norms and the set of all respondents (Chapter 6). Chapter 7 provides a detailed analysis of the effect of high maintenance assignments on the core job variables.

Appendix III, entitled "Research Cases," separates the data by organization. The survey instruments are described in more detail in other appendices. The bibliography completes the appendix section.

The researchers wish to express their gratitude to two organizations that provided partial funding for the project: Standard Oil of Indiana and the Federal Institute of Computer Technology. We are also appreciative of the important assistance of Joanne Colter and the excellent word processing work of Kathy Claybaugh.

J. Daniel Couger and Mel A. Colter
Colorado Springs

BIOGRAPHIES OF THE AUTHORS

J. Daniel Couger is Distinguished Professor of Computer and Management Science at the University of Colorado, Colorado Springs (UCCS). After 20 years in the technical area, Dr. Couger began his motivation research in 1977: That research has produced 11 papers, a book and a monograph. Dr. Couger was designated U.S. Computer Science Man-of-the-Year in 1977. He is listed in Who's Who in America and Who's Who in the World. He has served as a consultant to more than 30 organizations, including many Fortune 500 firms such as IBM, ITT, and GE. He was a manager in the computer field prior to his entry into academia.

Mel A. Colter is an Associate Professor of Information Systems and Management Science at the University of Colorado, Colorado Springs (UCCS). He has been involved with systems efforts since 1965 in both industry and academic positions. His primary interests lie in the effective application of systems concepts to the development and management of computer resources. Dr. Colter has directed systems development in industry before moving to academia. He has published several articles and a book (with J. D. Couger and R. Knapp) on analysis and development techniques. He lectures internationally and serves as a consultant to many organizations from both the public and private sectors. He is a member of several professional organizations and he is listed in Who's Who in the World.

SECTION 1:
MAINTENANCE PROJECT OVERVIEW

CHAPTER 1
THE MAINTENANCE PROBLEM

The principal purpose of this research project has been the analysis of the effect of maintenance assignments on the motivation of programmers and analysts. It is a continuation of the research begun by J. Daniel Couger and Robert A. Zawacki in 1978, culminating in their research report, Motivating and Managing Computer Personnel (Wiley-Interscience, 1980).

That work identified the key factors for motivating programmers and analysts and how individual characteristics of these personnel affected their response to their motivational environment.

This project concentrated on a specific and important area of task assignment for programmers and analysts—the task of maintaining existing systems.

That area of activity has been singled out for research because of two principal reasons:

1. For most organizations it occupies over half of the personnel budget for the analysis/programming function.
2. Maintenance work is generally perceived to be less interesting and resulting in higher employee turnover rates than new system development work.

Therefore, the maintenance activity is a problem area—both in terms of the type of work and the magnitude of budget occupied.

If the research could identify better ways to organize this work and better ways to assign tasks to personnel, the impact on the data processing industry would be substantial.

Fortunately, the research accomplished its objective. Substantive research outcomes were achieved which should benefit the computing community significantly.

The report will discuss research design, methodology, data analysis and conclusions. The remainder of Chapter 1 is devoted to substantiation of the relevance and magnitude of the maintenance problem. It also identifies the scope of the research project.

COST OF MAINTENANCE

Figure 1.1 provides data on maintenance cost compared to other hardware/software costs. It was developed by Barry W. Boehm, long recognized as an authority in software cost projection and software engineering techniques.[1] The proportion of the hardware/software cost allocated to system maintenance, in the typical U.S. computer installation, is not only more than half of the total software cost, it is also approximately 50 percent of the total computing cost.

A recent study by the U.S. General Accounting Office (GAO) estimated that the federal government spends at least $1.3 billion annually on software maintenance (excluding weapons system software).[2] Other findings of that study were:

1. Half of a programmer's time is devoted to maintenance.

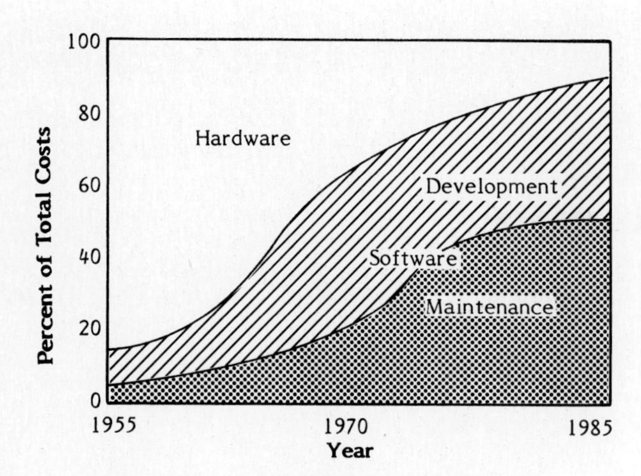

Figure 1.1: Hardware/Software Cost Trends (Boehm)

2. Some estimates say that up to 60% of all ADP dollars will be spent on software maintenance in the future if the present growth rate continues.
3. One recent DoD study showed that the cost of development for Air Force avionics software averaged about $75 per instruction while the cost of maintenance corrections of deployed software has ranged up to $4,000 per instruction.

For some organizations the cost of maintenance is even higher. Figure 1.2 shows that it occupies over 60 percent of ten-year life cycle costs at General Telephone and Electric and at General Motors. Elshoff[3] indicates that the figure for General Motors is about 75 percent and that GM is fairly typical of large business software activities.

Daly[4] indicates that about 60 percent of GTE's ten-year life cycle costs for real time software are devoted to maintenance. On the two Air Force command and control software systems shown in Figure 1.2, the maintenance cost was 67 and 72 percent.[5]

The previously cited study by Boehm reported the annual cost of software in the U.S. in 1980 at approximately 40 billion dollars, or about two percent of the Gross National Product. The number of applications installed has increased substantially, increasing the maintenance load. In addition, the average system life has increased from three years in 1960 to five in 1970 and eight in 1980.[6] The GAO study mentioned earlier estimates the current cumulative Federal investment in software at $25 billion.

Maintenance is labor-intensive. John Diebold projects that by 1985 the cost of hardware will be one-tenth of the 1979 rate and that labor costs will be twice as high as the 1979 rate.[7]

These figures confirm the significance of the maintenance activity throughout the United States—in government and in private industry. Let us now examine the productivity problem in the maintenance activity.

VIEW OF MAINTENANCE AS UNINTERESTING WORK

"Maintenance . . . tends to act as a millstone around the neck of the development organization."[8]

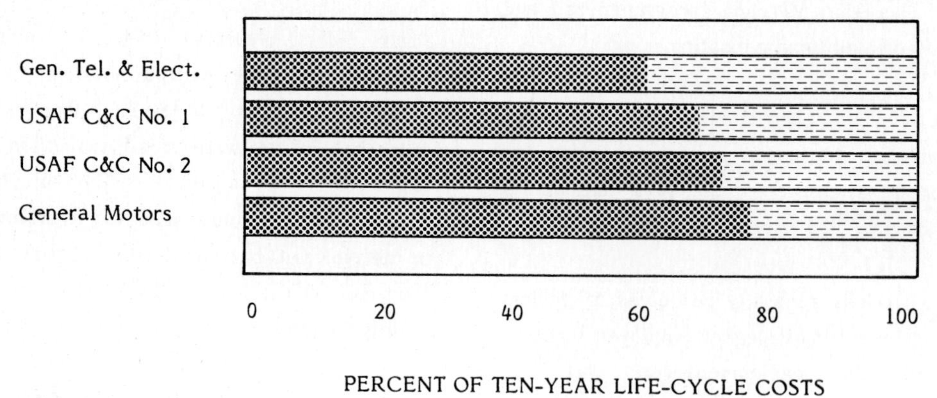

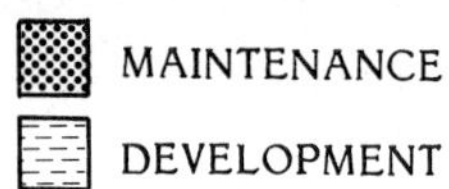

Figure 1.2: Maintenance Cost for G.M., G.T.E., and Two Large Air Force Systems (Boehm)

"Software maintenance tends to be viewed as an uninteresting, necessary evil . . ."
Richard G. Canning, editor, EDP Analyzer, August, 1981, p. 1.

* * * * * * * * * *

"Traditionally, program maintenance has been viewed as a second class activity, with an admixture of on-the-job training for beginners and of low-status assignments for outcasts and the fallen."
Richard E. Gunderman, "A Glimpse into Program Maintenance," Datamation, 1978, p. 99.

* * * * * * * * * *

"Analysts and programmers generally view the maintenance function as an inferior, noncreative, non-challenging activity."
Chester C. Lin, "A Look at Software Maintenance," Datamation, Nov., 1976, p. 5.

* * * * * * * * * *

"It is often difficult to get good people to want to work in a maintenance group."
Robert R. Jones, "Creativity Seen Vital Factor, Even in Maintenance Work," Computerworld, Jan. 23, 1978, p. 25.

* * * * * * * * * *

"Not only is the management of maintenance activities difficult for the data processing manager, but it is compounded by the fact that his staff is not generally interested, from a professional standpoint, in becoming maintenance programmers."
Bob McGregor, "Program Maintenance," Data Processing, May-June, 1973, p. 173.

Figure 1.3: Views on Maintenance Work

As the above quotation illustrates, maintenance is a problem area for management—not just because of the high cost—but also because of the way the work is perceived by computer professionals. Figure 1.3 provides some of the published views on maintenance. The literature is devoid of opposing opinions.

Gerald Weinberg tells a story about a programmer who frequently changes companies—every time after his new development project gets installed and he is assigned to maintain it. Weinberg observes that, though apocryphal, the story is representative of what is happening in industry.

A similar view is expressed by McGlinchey in the article, "What Do Programmers Want?"[9] There are two principal reasons for turnover, he says: 1) people quit "because they get stuck in a one-project job for which they are over-qualified," and 2) "pigeonholing . . . getting known as the payroll expert and not given the opportunity to do anything else." The most recent Datamation survey showed turnover quite high, 28 percent overall and 34 percent for application programmers.[10]

The U.S. economic decline has reduced turnover rates; nevertheless, turnover in the computer field is still high compared to other professions. Turnover is costly. Although statistics on the cost of turnover are not available for the computer field, data from other fields provides a basis for estimating turnover costs. The insurance industry has estimated the replacement cost of $6,000 for a claims investigator, $24,000 for a field examiner, $31,600 for a salesperson and $185,000 for a sales manager. These figures include the cost of hiring and training. The source is dated (1972), so these costs need to be increased to account for inflation.[11]

Another cost of turnover is absent from the literature. In replacing maintenance personnel, the learning curve for incoming personnel is significant, due to the following reasons: 1) documentation tends to be poorer on older systems, 2) these systems are less structured, 3) they are written in assembly language or early versions of a high-level language, and 4) few of the people who originally worked on the system are available, in case questions arise.

The difference in learning curves for people replacing new development personnel versus those replacing maintenance personnel is characterized in Figure 1.4.

However, turnover cost due to loss of personnel dissatisfied by maintenance assignments is negligible compared to the cost to maintain these systems by dissatisfied people. If the industry view of maintenance work, characterized by comments in Figure 1.3, is valid—over 50% of the analysis/programming work in a computer department is being performed by people who consider the work unchallenging—even boring. The resulting loss in productivity for the entire U.S. amounts to millions of dollars per year.

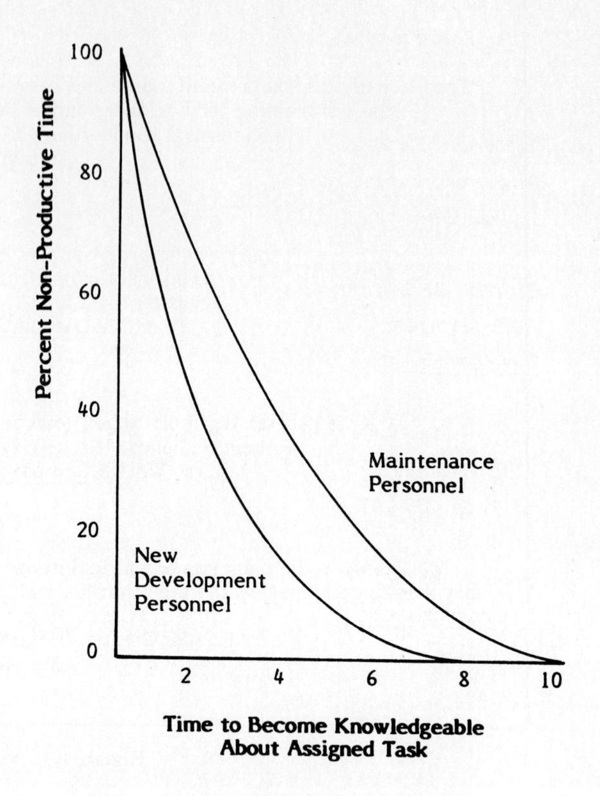

Figure 1.4: Learning Curve for Personnel, Development vs. Maintenance Projects (Empirically Derived)

POTENTIAL FOR IMPROVEMENT

Is Weinberg's apocryphal programmer representative of the industry? Perhaps so—for most persons in the field. However, contrary to the norm, our research identified persons who are both satisfied and productive in maintenance positions. Nor are these circumstances peculiar to certain companies or industries. The research identified approaches to insure a good motivational environment for almost every type of industry or governmental organization. More important, it identified ways to match people to tasks—depending on job and individual characteristics—to enhance the motivational potential.

One other aspect needs to be discussed prior to further explanation of our research project. Productivity of maintenance work can be improved two ways:

1. Improving the techniques for maintaining applications.
2. Improving the motivational environment for maintenance personnel.

This research project concentrated on the latter topic. However, we are continuing our work on a project covering improved techniques for analysis and programming in maintenance activities.

References for Chapter 1

1. Boehm, Barry W., Software Engineering Economics, 1981, Prentice-Hall, Inc., Englewood Cliffs, NJ, p. 18.

2. "Federal Agencies' Maintenance of Computer Programs: Expensive and Undermanaged," 1981, U.S. General Accounting Office, Gaithersburg, MD.

3. Elshoff, J. L., "An Analysis of Some Commercial PL/1 Programs," IEEE Transactions on Software Engineering, June, 1976, pp. 113-120.

4. Daly, E. B., "Management of Software Development," IEEE Transactions on Software Engineering, May, 1977, pp. 229-242.

5. Boehm, Barry W., "Software Maintenance," IEEE Transactions on Computers, December, 1976, pp. 1226-1241.

6. Lyons, M. J., "Structured Retrofit—1980," 1980 Proceedings of SHARE 55, pp. 263-265.

7. Diebold, J., "Improving the Utilization of Personnel Resources," The Diebold Computer Planning and Management Service, August, 1979, pp. 44-46.

8. Howard, Philip, "Examining the Maintenance Issue," System Development, April, 1982, p. 4.

9. "What Do Programmers Want?" Output, February, 1981, pp. 56-61.

10. McLaughlin, R. A., "That Old Bugaboo, Turnover," Datamation, October, 1979, p. 96.

11. Flamboltz, E. G., "Toward a Theory of Human Resource Value Accounting in Formal Organizations," The Accounting Review, Vol. 47, 1972, pp. 666-678.

CHAPTER 2
THE MODEL OF MOTIVATION

The C-Z diagnostic survey instrument was developed in 1978 by J. Daniel Couger and Robert A. Zawacki to establish motivational norms for jobs in the computer field. It was evolved from the Job Diagnostic Study instrument developed by J. Richard Hackman (University of Illinois) and Greg R. Oldham (Yale University), for two principal reasons:

1. The Hackman/Oldham instrument is conceptually sound. Its validity and reliability have been substantiated in studies on more than 500 different jobs in more than 50 different organizations.[1]

2. A major objective was to compare results with prior studies of personnel in other professions. Hypotheses on the difference between computer personnel and other personnel could be tested.

The expanded survey questionnaire includes other elements: employee perceptions on relative importance of problems relating to maintenance, realistic work schedules, access to the computer, access to supervisors and access to others (e.g., users or personnel in other departments whose work affects ours). This modified instrument is called the C-Z Survey.

The expanded instrument was then revalidated (in 1978), and administered to computer personnel representative of all industry segments. The Couger-Zawacki data base contains information on more than 6,000 persons in 18 job types in the computing field.[2]

MODEL OF MOTIVATION

The Hackman/Oldham JDS research was based on the model of motivation which evolved successively from the work of Abraham Maslow,[3] Frederick Herzberg,[4] and A. N. Turner and P. R. Lawrence.[5] Hackman and E. E. Lawler built on this prior research to isolate the key job dimensions to elicit motivation.[6] The Hackman/Lawler model of motivation (Figure 2.1) identifies the three "critical psychological states" associated with high levels of internal motivation, satisfaction, and quality of performance. These psychological states are 1) experienced meaningfulness of the job; 2) experienced responsibility for outcomes; and 3) knowledge of actual results. The existence of these psychological states in a job should lead to low absenteeism and turnover and high levels of internal motivation, satisfaction and quality of performance.

Five characteristics of the job (core job dimensions) are analyzed by the survey instrument: skill variety, task identity, task significance, autonomy, and feedback. Each core dimension is defined below:

1. Skill Variety: The degree to which a job requires a variety of different activities in carrying out the work, which involve the use of a number of different skills and talents of the employee.
2. Task Identity: The degree to which the job requires the completion of a "whole" and identifiable piece of work—i.e., doing a job from beginning to end with a visible outcome.

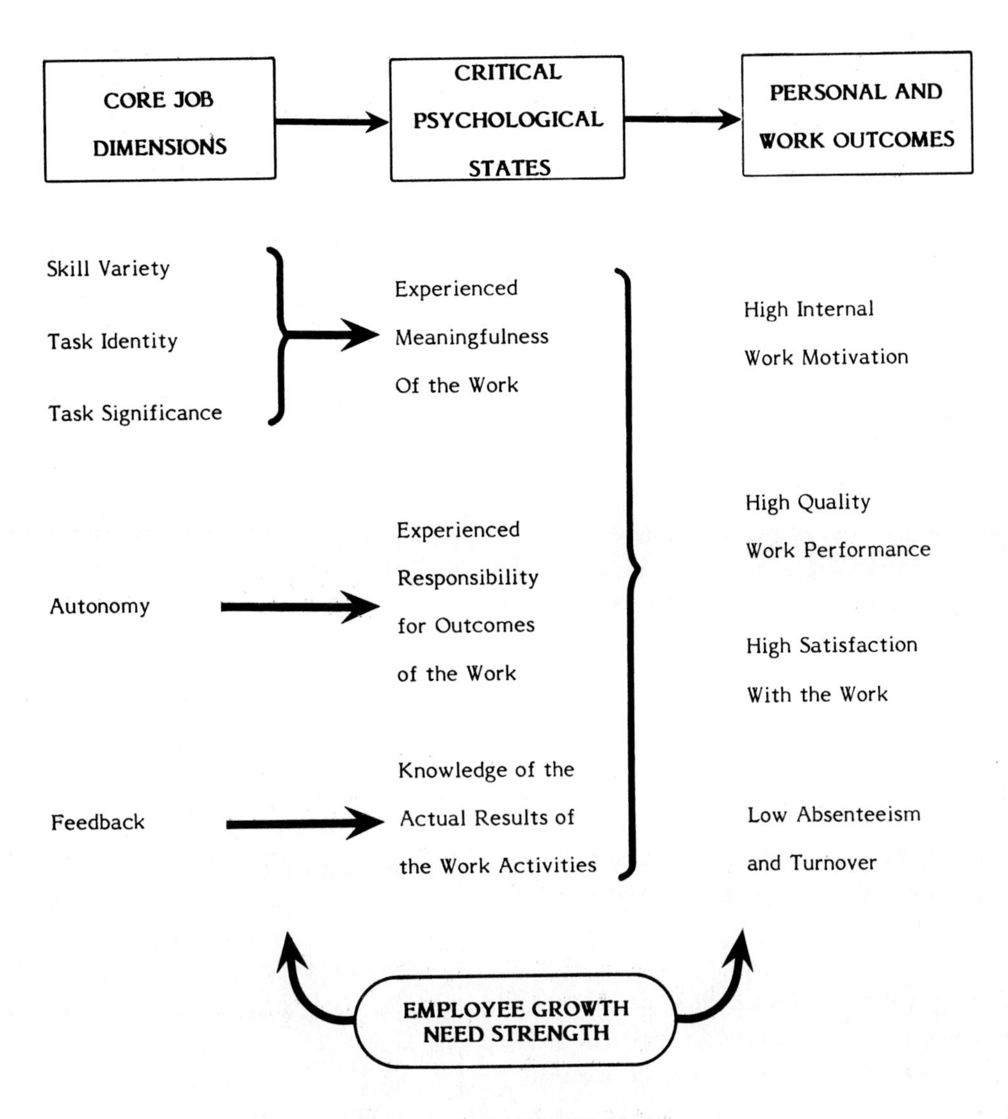

Figure 2.1: Model of Motivation (Source: Hackman and Lawler)

3. Task Significance: The degree to which the job has a substantial impact on the lives or work of other people—whether in the immediate organization or in the external environment.
4. Autonomy: The degree to which the job provides substantial freedom, independence, and discretion to the employee in scheduling his/her work and in determining the procedures to be used in carrying it out.
5. Feedback from the Job Itself: The degree to which carrying out the work activities required by the job results in the employee obtaining information about the effectiveness of his or her performance.

THE EFFECT OF PROPER LEVEL OF CORE DIMENSIONS

Hackman and Oldham use the three "psychological states" experienced by a golfer to illustrate the effect of core dimensions. "Consider, for example, a golfer at a driving range, practicing to get rid of a hook. His activity is meaningful to him; he has chosen to do it because he gets a 'kick' from testing his skills by playing the game. He knows that he alone is responsible for what happens when he hits the ball. And he has knowledge of the results within a few seconds."[7] The psychological states are defined as follows:

1. Experienced Meaningfulness: Individuals must perceive their work as worthwhile or important by some system of values they accept.
2. Experienced Responsibility: They must believe that they personally are accountable for the outcomes of their efforts.
3. Knowledge of Results: They must be able to determine, on some fairly regular basis, whether or not the outcome of their work is satisfactory.

If these conditions exist, people tend to feel very good about themselves when they perform well. Those good feelings motivate them to try to continue to do well. This is what the behavioral scientists mean by "internal motivation"—rather than being dependent on external factors (such as incentive pay or compliments from the boss). The JDS model computes a single summary index which indicates the "motivating potential" of a job. The index is called the motivating potential score (MPS).

When all three psychological states are high, then internal work motivation, job satisfaction and work quality are high and absenteeism and turnover are low. A number of studies by Hackman and Oldham have validated this conceptual model.

Table 2.1 provides the results (on a scale of seven) from the Couger-Zawacki national survey of analysts and programmers. These are the norms, not the national average. The organizations included in the survey were carefully selected to insure that they had a satisfactory motivational environment. The concept of the norm is further discussed on page 15.

Core Job Dimensions	Programmers and Analysts
Skill Variety	5.408
Task Identity	5.205
Task Significance	5.605
Autonomy	5.290
Feedback From Job	5.113

Table 2.1: National Norms for Core Job Dimensions (Source: Couger and Zawacki)

Job Category	Growth Need Strength	Motivating Potential Score
DP Professionals	5.91	153.6
Other Professionals	5.59	153.7
Sales	5.70	146.0
Service	5.38	151.7
Managerial	5.30	155.9
Clerical	4.95	105.9
Machine Trades	4.82	135.8
Bench Work	4.88	109.8
Processing	4.57	105.1
Structural Work	4.54	140.6

Table 2.2: Comparison of GNS and MPS by Job Category (Source: Couger and Zawacki)

INDIVIDUAL'S GROWTH NEED STRENGTH MATCHED AGAINST JOB'S MOTIVATING POTENTIAL

The expectation is that people who have a high need for personal growth and development will respond more positively to a job high in motivating potential than people with low growth need strength.

Obviously, not everyone is able to become internally motivated—even when the motivating potential of the job is quite high.

Behavioral research has shown that the psychological needs of people determine who can (and who cannot) become internally motivated at work. Some people have a strong need for personal accomplishment—for learning and developing beyond where they are now, for being stimulated and challenged, and so on. These people are high in "growth need strength" (GNS). Figure 2.2 diagrammatically shows how individual growth needs affect response to a job high in motivating potential.[8]

The key reason for computing growth need strength is to compare it with the job's potential to fulfill an employee's need for growth. A job low in motivating potential will

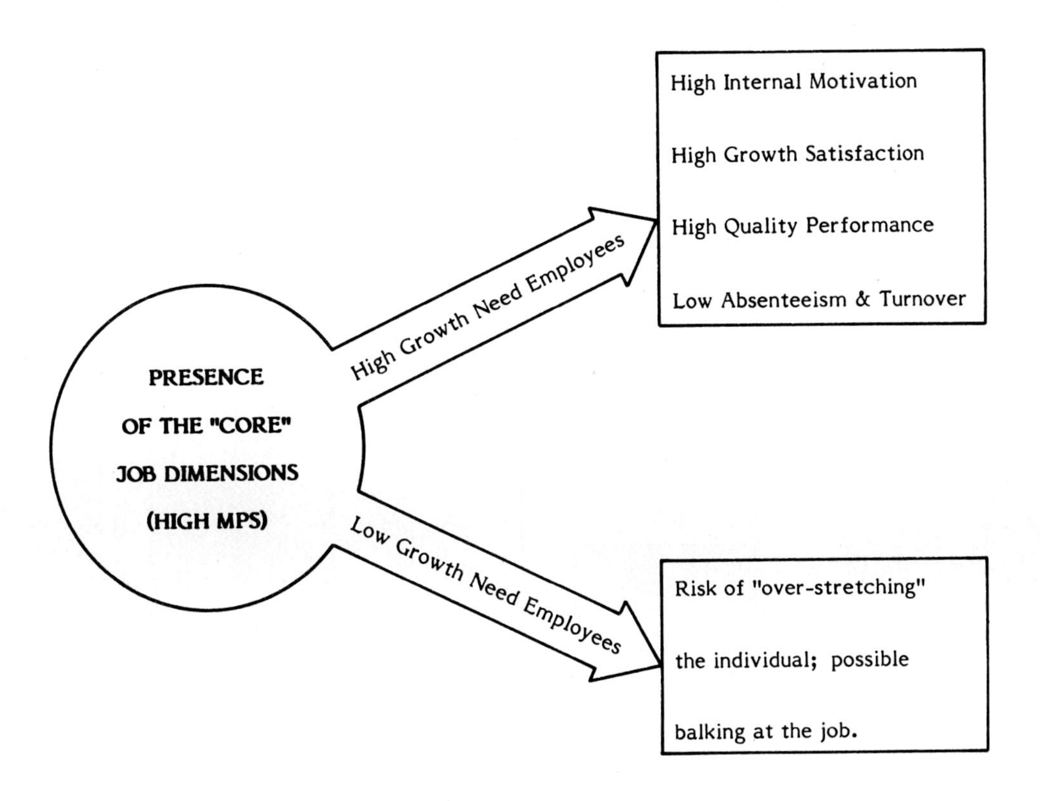

Figure 2.2: The Effect of a High Motivating Job On Persons With Varying Growth Need Strength (Hackman and Oldham)

frustrate a person with high growth need strength. It is a perfect example of the old cliche of a square peg in a round hole.

Growth need strength is quite high for computer professionals, compared to other professionals and to other job categories. The left-hand column of Table 2.2 shows this effect (responses are on a scale of seven). Table 2.2 lists GNS for a wide range of occupations—the highest to the lowest of the 500 occupations measured by Hackman and Oldham.

The MPS, shown in the right-hand column of Table 2.2, enables a comparison of employee growth need strength and the job's potential to motivate that person. MPS is a computed value (not on a scale of seven), reflecting the potential of a job for eliciting positive internal work motivation on the part of employees. The Hackman/Oldham survey results provide an illustration of the imbalance of GNS and MPS. Notice in Table 2.2 that the lowest GNS is for structural work. On the other hand, the MPS for that field is 140, which is near the midpoint for all jobs reported in the table. Jobs in that industry have a motivating potential above the growth needs of the workers. The lower prong on the two-arrow diagram in Figure 2.2 illustrates this situation. The result is a mismatch of GNS and MPS. The opposite situation is illustrated by the data on clerical workers. Note that GNS of these personnel are near the middle of the GNS array, 4.95. MPS for these personnel is quite low, 105.9. This mismatch on GNS and MPS will result in poor motivation as will the mismatch for structural workers.

On the other hand, the upper prong on Figure 2.2 illustrates the healthy situation for the national norms for computer professionals. Both the individual's growth need strength and the job's motivating potential score are high. A good GNS/MPS match can be expected to produce high levels of motivation.

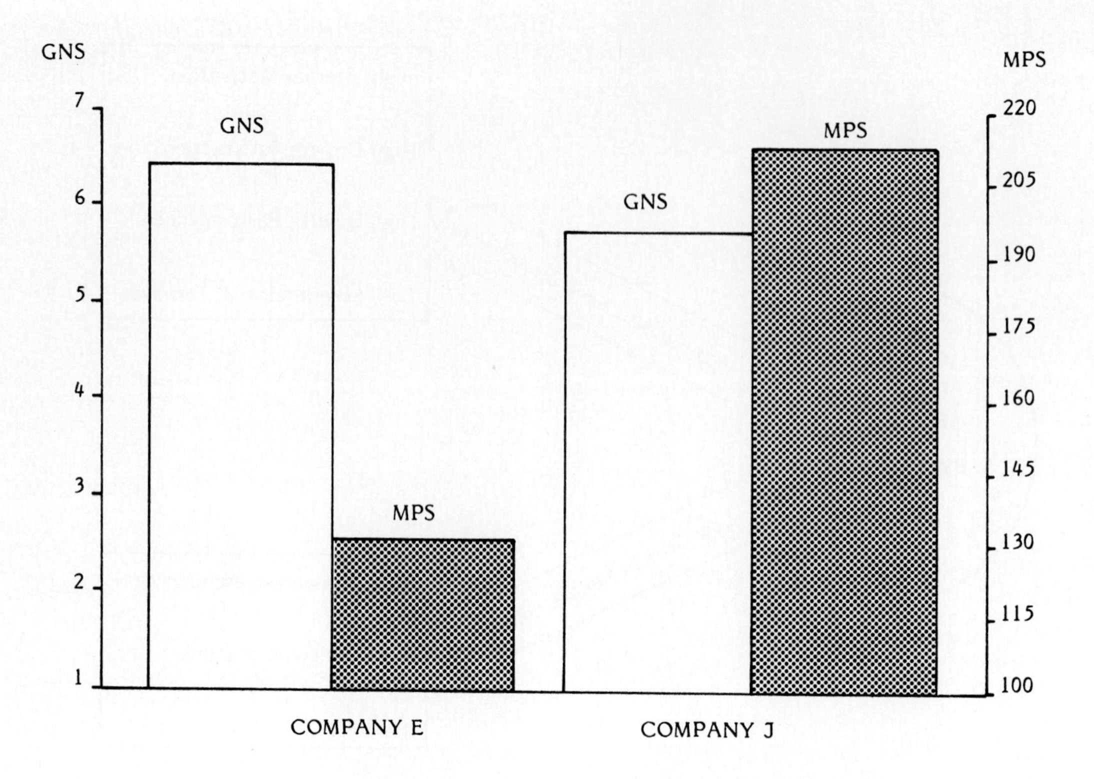

Figure 2.3: Imbalance of GNS/MPS in Two Organizations, Resulting in Poor Motivational Environments (Couger and Zawacki)

The goal is a match in MPS and GNS. If both are equally low, a balance exists and motivation will result. Likewise, high GNS and MPS will produce a match that results in motivation. Lack of motivation results when a mismatch occurs. The structural and clerical jobs described above illustrate this problem. Mismatches occur in our field too, as discussed next.

MISMATCH IN MPS/GNS = LOW MOTIVATION

Although the Couger-Zawacki national norms show a high positive correlation between GNS and MPS, this is not necessarily the case for individual organizations. For example, one of the organizations in our earlier research had higher than normal GNS (6.20) but one of the lowest MPS (130.2) (see left side of Figure 2.3). In contrast, an organization with low GNS (5.73) had the highest MPS (209.8) of any organization studied (right side of Figure 2.3). Both these organizations exhibited a mismatch between GNS and MPS and, hence, a poor motivational environment. Accordingly, productivity was low.

It can be seen by these examples that the C-Z survey instrument is an important device to measure individual GNS versus the job's MPS. In the current research project, the C-Z survey provided the means to measure the impact of maintenance assignments on the job's MPS. It also provided the means to measure the GNS of persons assigned to maintenance. The degree of match in MPS/GNS could be ascertained.

IMPROVING GNS/MPS MATCH-UP

Once the mismatch in GNS/MPS is identified, corrective action can be undertaken. The firms in Figure 2.3 will be used as an example.

Company E. The firm depicted by the two columns on the left had a motivationally deficient programmer job, where MPS was significantly below the national norm. Conversely, mean GNS for programmers in this organization was significantly higher than the national norm. Figure 2.4 restates the data in Figure 2.3 showing

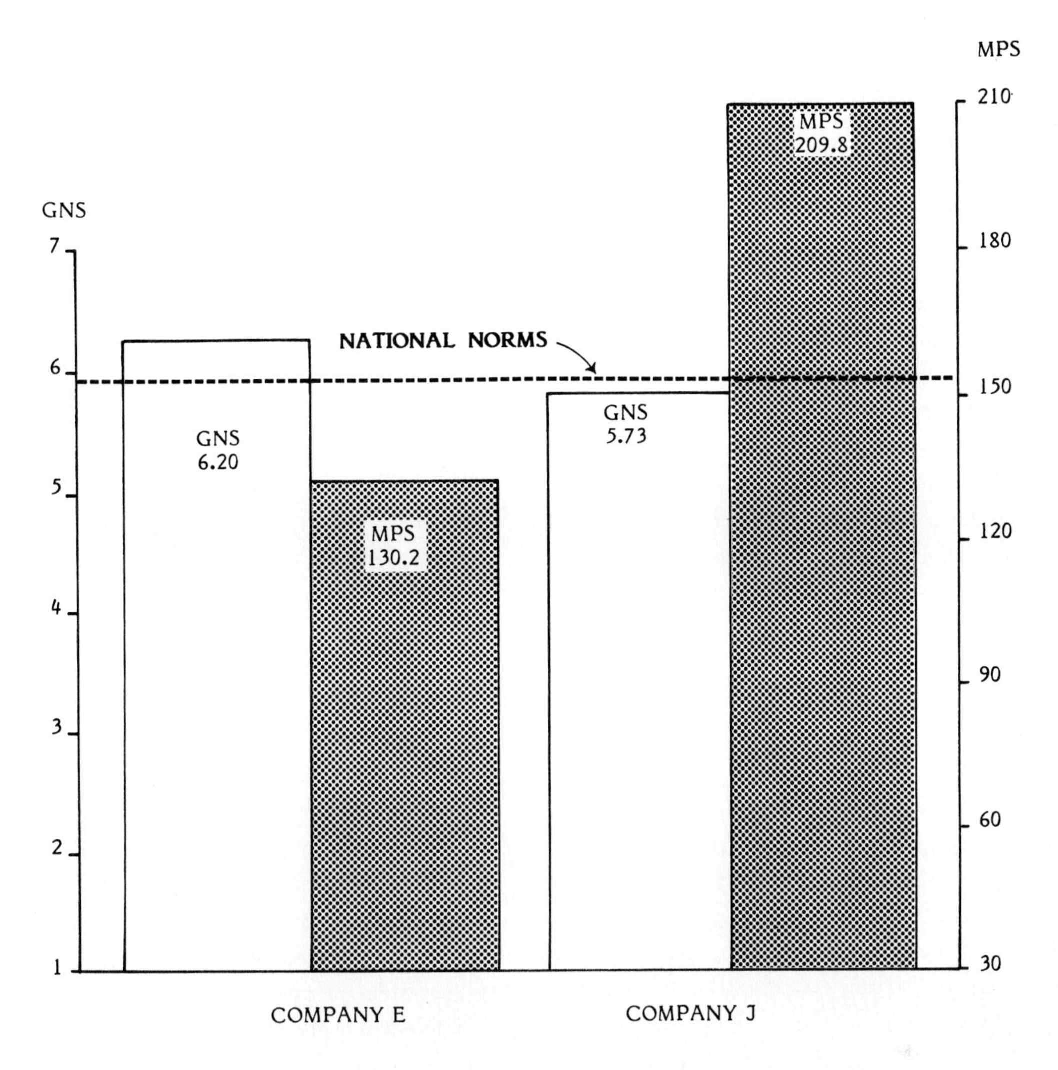

Figure 2.4: Two Companies Shown in Figure 2.3 Compared To National Norms. For Both Companies, MPS and GNS Are Significantly Different From The Norms.

employee GNS and job MPS compared to the national norm. The horizontal bar at the top of the chart represents the national norm for both GNS and MPS. GNS is on the left scale and MPS on the right scale.

Investigation of the cause for this discrepancy revealed that the company had undergone major changes in the previous two years. A new Vice-President of Data Processing replaced one whose performance was sub par. The new V.P. was given a substantial budget increase "to bring the organization's computer applications to a level of sophistication consistent with competitors." He hired experienced top performers. These proven, high energy personnel raised the mean GNS well above the national norm. Nevertheless, the programming job was perceived to be very narrow and unchallenging. The huge gap in MPS and GNS was substantiated by low productivity.

Analysis of the programming job to determine the cause for low MPS showed that the problem concerned the company's approach to implementation of the Chief Programmer Team Concept. The employees referred to the five chief programmers as "superprogrammers." Management let these high-output people do nothing but program design and coding. The rest of the team was used to relieve the chief programmers of testing and documentation functions and any other activity which might lessen lines-of-code-produced.

The result was a very high MPS job for five persons and

low MPS jobs for the remainder of the programming staff. The feeling of the latter group was summed up by the comment of one, "we're here only to support the superstars—we'd like to have a piece of the action." Consequently, the mean MPS was substantially below the national norm. The MPS/GNS mismatch was resolved as follows. The policy of relief from mundane activities for the five chief programmers was not changed, in order to maintain their high level of productivity. Personnel were transferred from operations and were trained to handle the testing and documentation activities. The unhappy intermediate level programmers were reassigned to projects where their responsibilities included program design and coding. This raised MPS for the programming group without lowering that of the chief programmers.

A longitudinal study (re-survey) six months later showed that MPS had been raised slightly above the national norm. The resulting GNS/MPS match increased motivation and, concomitantly, productivity.

Company J. The situation for the company depicted by the two columns on the right was the reciprocal of Company E. GNS was significantly lower than the national norm and the MPS was significantly higher. Although the factors were reversed, the result was similar to that of Company E—poor GNS/MPS match and low productivity.

Investigation revealed very different causes from the Company E situation. Company J had a standard promotion path from operations to programming. The national survey showed that GNS for operations personnel is significantly lower than that of programmers. A company with a standard promotion path from operations to programming could expect to have mean programmer GNS below the national norm. Surprisingly, this company had the richest job of any company we've studied. Examination of the job identified the reasons. Here there were no programmers, per se; all personnel were programmer/analysts. They interviewed users to establish requirements; developed system specifications; designed the system and the programs; coded, debugged and tested programs; conducted system tests; wrote operation and user procedures; documented the system; trained users and implemented the system. They performed every step in the system development cycle. The job was surpremely rich. Yet it was filled by persons whose GNS called for a job less rich than the average programmer's job.

In the case of Company E, we recommended enhancement of the jobs to raise MPS to the level consistent with the employees' GNS. The programmer jobs were enlarged to that of a programmer/analyst. In Company J, we recommended just the opposite—reducing the scope of activity and permitting specialization—separating analyst and programmer jobs. The longitudinal study on this company showed a match in GNS/MPS. Motivation and job satisfaction were high. Productivity has increased substantially.

Work Redesign Versus Task Reallocation.

The two examples above represent what the behavioral scientists refer to as work redesign. Another approach, equally effective, but much simpler to accomplish, is task reallocation. When the differential between GNS and MPS is less pronounced, supervisors can allocate tasks, or make application assignments, according to the richness desired by the individual. Lower GNS personnel are assigned smaller scope jobs, i.e., an enhancement rather than a new system development. Or they are assigned less complicated applications, i.e., the check writing subsystem of payroll system instead of the scheduling subsystem of the MRP system.

Procedure to Analyze Jobs to Modify MPS.

The above examples provided an overview of an MPS change process. The detailed procedure involves analysis of the individual five core job dimensions. Some job dimensions may be required to be held constant due to policy or procedural constraints. The process of analyzing individual core job dimensions is referred to as motivational analysis. It will be explained and illustrated in detail in Chapter 4. In this chapter the intent is to impart an understanding of MPS and GNS and how they are varied to produce a match, insuring motivation. The key point is the match. A lower GNS person whose job has a matching MPS can be expected to be motivated, and hence, productive. The same is true for a high-GNS person.

Changing GNS Versus Changing MPS

All of the above examples dealt with modifying MPS to meet an individual's or a group's MPS. Another alternative is varying GNS. If a supervisor has available a rich assignment, such as data base design, it can be assigned to a high-GNS person.

However, a person's GNS rarely changes. The behavioral scientists have shown that GNS is an acquired trait—typically very early in our lives. Therefore, in matching tasks to individuals, supervisors are constrained. They can vary the task assignments for an individual but not his/her GNS. Therefore, in the examples in Figure 2.3 and 2.4, we worked on changing the job to produce a GNS/MPS match.

A drastic revision in company GNS is obtained only by wholesale firing and replacement. Obviously, this is not a viable alternative. In the long run, mean GNS can be modified by changes in selection procedure. For example, Company E raised its mean GNS by adding a number of high-GNS persons over a six-month period. Management of Company J also raised its mean GNS by changing the policy of promotion from operations to programming. It changed it from a standard path to a selective path, where those persons with the desired aptitude and GNS were given the opportunity to move to programming. Management was also concerned with equity and elevated the career path within operations to enable persons to select either of the two career paths, depending on company need and individual qualifications.

In Chapter 4, we will provide procedures for assessing an individual's GNS and a job's MPS. With these procedures a supervisor can work on improving the GNS/MPS match-up.

These procedures are especially effective for enhancing maintenance jobs. The research has shown that the general model of motivation can be applied to the computer field with substantial improvement in motivation.

References for Chapter 2

1. Hackman, J. R. and G. R. Oldham, "Development of the Job Diagnostic Survey," Journal of Applied Psychology, Vol. 60, No. 2, 1975, pp. 159-170.

2. Couger, J. Daniel and Robert A. Zawacki, Motivating and Managing Computer Personnel, 1980, Wiley-Interscience.

3. Maslow, Abraham, Motivation and Personality, New York: Harper & Row, 1954.

4. Herzberg, Frederick, The Motivation To Work, John Wiley & Sons, Inc., New York, 1959.

5. Turner, A. N., and P. R. Lawrence, Industrial Jobs and the Worker, Harvard Graduate School of Business Administration, 1965.

6. Hackman, J. R. and E. E. Lawler, "Employee Reactions to Job Characteristics," Journal of Applied Psychology Monograph, 1971, pp. 259-286.

7. Hackman, J. R., G. R. Oldham, R. Janson, and K. Purdy, "A New Strategy for Job Enrichment," California Management Review, Vol. 17, No. 4, 1975, p. 58.

8. Ibid., p. 60.

Definition of the Word "Norm"

The term "norm" is used rather than "average" in this book because the companies chosen to be included in the survey had "above-average" motivational environments. They were selected because they had satisfactory motivating environments - without significant problems. Since a number of U.S. companies have motivational problems, a national average is inappropriate as a target for improvement. In other words, these are "standards" for motivation. The standards, or norms, are what is being accomplished in those organizations with "healthy" motivational environments. It is possible for any organization with average resource levels to achieve the norms.

CHAPTER 3
THE RESEARCH METHODOLOGY

Two research instruments were used for this project: 1) the Couger-Zawacki Diagnostic Survey (C-Z Survey) and 2) the Structured Interview Questionnaire on Motivation (SIQM). They will be explained before discussing the firms involved in the research.

COUGER-ZAWACKI SURVEY

The C-Z survey is administered individually to each employee. It takes 25 minutes to complete. The survey instrument measures: 1) job dimensions, 2) satisfaction levels, 3) goal clarity/accomplishment, 4) DP problem areas. Most important, it measures the potential of each job to motivate the specific group of employees holding that job. When a mismatch occurs, clues are provided on how to realign the task assignments.

Two types of reporting occur: 1) comparison to national norms by job type, 2) comparison to national norms by organizational unit. The C-Z survey diagnoses which jobs have low motivating potential and compares MPS to the GNS of persons filling that job.

In addition to the seven variables defined in Chapter 2, the survey provides the following:

1. Satisfaction Measures: The private, affective reactions or feelings an employee gets from working on his job.
   - A. General Satisfaction: An overall measure of the degree to which the employee is satisfied and happy in his or her work.
   - B. Internal Work Motivation: The degree to which the employee is self-motivated to perform effectively on the job.
   - C. Specific Satisfactions: These scales tap several specific aspects of the employee's job satisfaction.
     1. Pay satisfaction
     2. Supervisory satisfaction
     3. Satisfaction with co-workers
2. Social Need Strength: This is a measure of the degree to which the employee needs to interact with other employees.
3. Goal Clarity and Accomplishment: These scales measure the degree to which employees understand and accept organizational goals. Further, they tap into the employees' feelings about goal setting participation, goal difficulty, and feedback on goal accomplishment.
4. Problem Areas: This is a measure of the degree to which the following areas are problematic:
   - A. Amount of maintenance being performed
   - B. Access to the computer
   - C. Realistic schedules
5. Experienced Meaningfulness: This scale is a measure of how worthwhile or important the work is to the employees.
6. Experienced Responsibility: This scale measures the employees' beliefs that they are personally accountable for the outcomes of their efforts.
7. Knowledge of Results: This scale measures the employees' beliefs that they can determine, on some fairly regular basis, whether the outcomes

of their work are satisfactory.

8. Individual Growth Need Strength: This scale measures the individual's need for personal accomplishment and for learning and developing beyond his/her present level of knowledge and skills.
9. Motivating Potential Score: A score reflecting the potential of a job for eliciting positive internal work motivation on the part of employees.

The firms involved in the research project were first sent a package containing the questionnaires and answer sheets. All of the data processing employees of the firms were encouraged to participate in the survey, though only those responses from programmers, analysts, and programmer/analysts were included in the evaluation of the results for this report. Data on other jobs was provided to management to facilitate motivation analysis of the entire organization.

Following the analysis of the C-Z Survey data, each firm received a two day on-site visit by at least one of the two researchers. During that visit, top level DP management was apprised of the results of the C-Z Survey study. Then, in a meeting with first line DP managers, the early results were again explained, and the structured interview questionnaire was used to elicit information from that group about the maintenance process in general for the firm. The remainder of the two days was spent in a series of one to 1½ hour interviews with employees.

STRUCTURED INTERVIEW QUESTIONNAIRE

While the C-Z Survey provides a measure of an individual's perceptions about his/her work environment, it pinpoints effects rather than causes. For example, if the survey results showed MPS to be lower than the national norms for DP, we could pinpoint the principal problem areas. It might be lack of skill variety or autonomy or task identity or feedback or task significance—any or all of these.

Our objective was to go further—to identify root causes, e.g., why was skill variety low. Therefore, a second level of research was designed to get at these root causes. The structured interview questionnaire was developed for that purpose. Appendix II provides a copy of the instrument. It covers three subject areas:

Part I

1. Questions on interviewee's perceptions on the reasons behind his/her unit's C-Z Survey responses.
2. Questions about interviewee's opinions on his/her own responses on the key job dimensions.

Part II

Questions on role clarity, role conflict and clarity of reward linkages.

Part III

Questions on interviewee's perceptions about maintenance— definition, problems and performance evaluation criteria.

The interview questionnaire was validated on two firms before the main survey was undertaken.

SURVEY DESIGN

The criteria for selection of firms to be included in the research project were as follows:

1. Representative of both public and private organizations.
2. Representative of centralized and decentralized data processing organization and equipment.
3. Representative in the mix of maintenance/new development work.

DESCRIPTION OF SURVEY PARTICIPANTS

Ten organizations, geographically dispersed, were included in the survey:

1. A city government organization
2. A state government organization
3. A federal government organization
4. A software development firm
5. An energy supplier (primarily oil)
6. Three consumer product suppliers/retailers
7. A government contractor
8. A regional center for one of the above

The organizations ranged in size as follows:

1. Four small-sized organizations—25 to 50 analysts/programmers.
2. Three medium-sized organizations—50 to 100 analysts/programmers.

3. Three large-sized organizations—100 to 200 analysts/programmers.

A total of 555 usable responses were obtained. Interviews were conducted with a sample of 104 persons, which included 43 supervisors. The interviews were conducted with personnel whose assignments ranged from zero to 100 percent maintenance.

The C-Z Survey was conducted in the firms; then, results were compiled and analyzed before the interviews were conducted. That way the researchers had in-depth knowledge about the firm's motivational environment (by job type and by organizational unit) before the interviews.

Data from the individual interviews were tied back to the C-Z Survey data for these persons by a code. The code—mother's maiden name initials—allowed precise identification of individuals but preserved the anonymity of the individuals from supervisory cognizance. Managers were also interviewed (in a group) in each firm—primarily, to better understand the work activity and approach to maintenance.

The types of data gathered fell into four broad classes. First, data on a set of job/individual variables were derived from use of the C-Z Survey instrument. From the same instrument, a second category of data was acquired: demographic data. The set of demographic variables is:

- -Sex
- -Age
- -Education
- -Marital Status
- -Years With Firm

The variable 'Number of Dependents' was also collected, as it has been found to hold more predictive power than 'Marital Status.' However, the variable 'Marital Status' will be used in this paper to ensure consistency with the national results. In addition, the respondent's estimate of the amount of his or her time spent on software maintenance activities was collected as a percentage.

The job/individual variables which were included in the analysis were discussed previously. Sample output of the C-Z Survey is provided in Appendix I.

The third category of data came from the structured portion of the interview questions. It included:

- -% Time on Maintenance Activities
- -% Time on Fixit Activities
- -% Time on Enhancement Activities
- -Definition of Maintenance (Fixit vs. Enhancement)
- -Average Maintenance Task Size in Days
- -Average Maintenance Task Size in Lines of Code (LOC)
- -Relative Severity of Maintenance Problems, broken into:
  - -Poor Documentation
  - -Poor Existing Design
  - -Poor Existing Code
  - -Unrealistic Schedules
  - -Other Problems
- -Relative Importance of Individual Performance Evaluation on Maintenance Tasks:
  - -Lines of Code Produced
  - -Lines of Error-Free Code Produced
  - -Compliance with Schedules
  - -Minimizing Costs
  - -Other Areas

The fourth category of data gathering concerned role clarity, role conflict, and clarity of reward linkages (to performance).

Note that the percent of time spent on maintenance was collected both on the C-Z Survey instrument and during the early portion of the interview. In both of these cases, the question was asked with no guidance as to what the researchers believe maintenance to be. Each respondent was allowed to answer the questions using whatever perception of the term he or she held. Then, in the interview process, the discussion of two components of maintenance was initiated. In all cases, the subjects agreed (and in many cases offered the information) that what is commonly called maintenance really consists of two components: those of 'fixit' and 'enhancement'. Fixit activities relate to the classic processes of expending resources on existing code because it has failed to work properly under some circumstances. Enhancement activities involve the wide range of efforts which are expended in order to keep programs current and expand their capabilities. Other than this general distinction, no firm or individual involved in the process defined any clear line of demarcation between enhancement and new development.

The remaining questions in the interview process, involving maintenance task size, maintenance problems, and performance evaluation criteria, were directed towards an understanding of the components of maintenance and its importance in that organization.

DATA ANALYSIS PROCEDURES

The analysis of data was accomplished in several phases. First, the C-Z Survey data from all respondents from all firms was compared to the national averages for both demographics and job variables in order to test the hypothesis that the sample for this study came from the same population as that of the national data base established by Couger and Zawacki. Next, each firm was subjected to the same test. Due to the differences in delineating components of the maintenance process, it was necessary to define any firm-specific characteristics which may impact on the results of the maintenance questions. Third, the data from the interviewed subjects was compared, as a group, to both the national averages and the data from all respondents from all of the involved firms, for the reasons listed above. This set of analyses provided a baseline for the remainder of the study by establishing similarities and/or differences between the various data sets.

The next major level of analysis involved the examination of the effect of the relative maintenance load on the demographic and job variables. This evaluation was done first by comparing the national data base averages with those obtained from this research. Second, the effect of maintenance on the demographics and job variables was examined by comparing the values for each firm with those of its interviewed subjects. In this case, the analysis is not as rigorous as in the prior analyses due to the small sample sizes involved.

Finally, the interview questionnaire data was analyzed, mostly through graphical analysis because of the difficulty in applying statistical tests to the very small sample sizes involved at this level.

In all of the above comparisons, both means and variances were tested for significant differences. For the tests on variances, the standard F statistic was used. For the tests on the means, the assumptions of normality and independence were made. However, the assumptions of equal variances and equal sample sizes were clearly not possible. Therefore, the formulation of the T statistic and its associated degrees of freedom was chosen to allow for the effects of the widely differing sample sizes and variances. When both n_1 and n_2 were large, the Normal distribution was used instead of the T distribution.

The following T statistic was used,[1] where

$$t = \frac{M_1 - M_2}{\sqrt{\frac{S_1^2}{n_1} + \frac{S_2^2}{n_2}}}$$

In addition, the number of degrees of freedom for the T statistic was calculated to account for these problems, where

$$\nu = \frac{(\text{est } \sigma_{m_1}^2 + \text{est } \sigma_{m_2}^2)^2}{(\text{est } \sigma_{m_1}^2)^2/(n_1+1) + (\text{est } \sigma_{m_2}^2)^2/(n_2+1)} - 2$$

where

$$\text{est } \sigma_{m_1}^2 = \frac{S_1^2}{n_1}$$

and

$$\text{est } \sigma_{m_2}^2 = \frac{S_2^2}{n_2}$$

In addition to the tests on means and variances, correlation coefficients were calculated between the core job dimension variables and the level of maintenance activity. Also, regression runs were performed in order to investigate the predictive effects, if any, of the percent of time spent on maintenance on the core job dimension variables. These results will be discussed later.

SUMMARY OF THE RESEARCH METHODOLOGY

The research utilized the C-Z Survey instrument, validated previously in the Couger-Zawacki national study of motivational norms for computer personnel. Using this approach the researchers could compare results in the survey firms to the national norms. Ten firms and 555 persons were involved in the survey.

At the next level of analysis, the structured interview questionnaire on motivation (SIQM) was utilized. The researchers personally interviewed 104 analysts, programmers, and supervisors.

Detail of discussion of the data analysis will be provided in Section 4 and Appendix IV. A management summary of the analysis and recommendations are provided prior to the detailed data analysis.

Reference for Chapter 3

1. Nie, N. H. et al., Statistical Package for the Social Sciences, McGraw-Hill, 1975, pp. 269-270.

SECTION 2:
MANAGEMENT SUMMARY

CHAPTER 4
CONCLUSIONS & RECOMMENDATIONS

The survey produced some data which are highly important to management in this era of high-maintenance activity. First we will examine survey results that supported previous intuitive views and provided insight in areas which need more attention. Then, recommendations will be made to enhance the motivational environment for the entire maintenance activity.

LOW MPS FOR MAINTENANCE ACTIVITY IN GENERAL

In Chapter 1, a number of derogatory comments about maintenance work were cited from the literature. The question was raised concerning the validity of these negative views. The research provides credence for these views, as shown in Figure 4.1. These graphs convey the most important facts revealed in the research. The percent of maintenance is inversely correlated with the amount of the core job dimensions perceived to be present in the job. The higher the percent of maintenance, the lower the motivating potential of the job. According to Figure 4.1, motivation is extremely low for the typical DP organization where maintenance comprises a large portion of the workload. An enormous potential for productivity improvement is indicated by these figures.

On the other hand, the graph indicates that organizations with less than 20% of their work comprised of maintenance could give everyone "a dose" of maintenance work such that the relative productivity impact is lessened. However, the figures cited in Chapter 1 showed that most organizations have a maintenance workload above the 50% level. It also showed that the proportion of workload devoted to maintenance is growing every year. Therefore, the MPS would be reduced from 159 for the less than 20% level to 133 for the 41-60% level of maintenance activity, as shown in Figure 4.1. For persons involved in fixit activities the impact is even more detrimental. MPS drops from 159 (less than 20% fixit) to 99 (41 to 60% fixit).

What can management do in such an environment? Is maintenance just a necessary evil, or is there a way to reverse the severely negative attitude about maintenance work?

POTENTIAL FOR IMPROVEMENT

Some organizations have attempted to lessen the negative attitude about maintenance by renaming the activity. One organization insists that all employees refer to maintenance as "retrofit activity." Several other organizations entitled the activity as enhancement.

Such a change is merely a superficial approach to correcting the problem. Just as the janitorial field learned, if you change the name of a dull and boring job, you do not change the job content—it remains a dull and boring job. If you are in maintenance, a new title may help increase status when explaining your job to a person outside the company, but it does not make your work any more motivating.

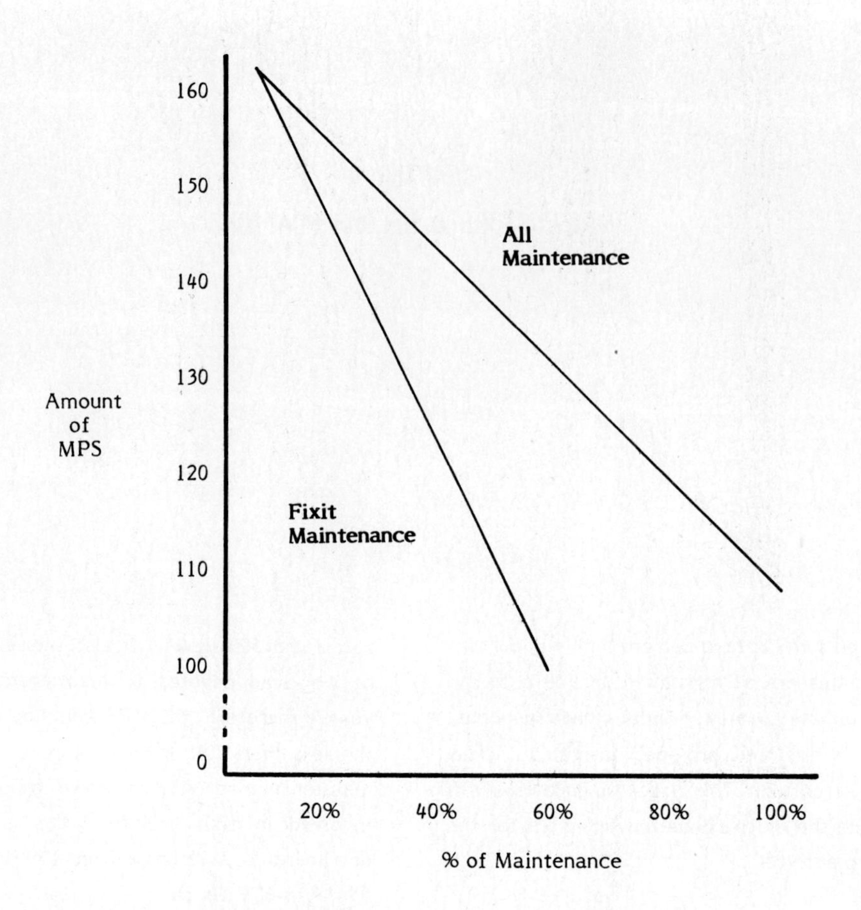

Figure 4.1: Effect of Percent of Maintenance On Job's Motivating Potential

Although a name change will produce little positive effect, there is promise for improving the situation. Analysis of the C-Z Survey results for individuals, supplemented by the interview process, revealed some individuals who were not de-motivated by maintenance work and who were considered by their supervisors to be quite productive. Some individuals were identified who believed their maintenance assignments to be challenging and interesting.

The research has enabled us to identify the job characteristics and the personnel characteristics of a productive maintenance environment so that other organizations can emulate these situations. To better take advantage of that potential, the reader needs some further explanation of the GNS/MPS match introduced in Chapter 2.

GNS/MPS MATCH

Previous motivation research showed the degree of "richness of a job" (its MPS level) needs to be matched with the individual's need for a rich job. Not all individuals are seeking enrichment in their jobs. The individual characteristic that is most important in the MPS/employee match-up is growth need strength. High GNS people are seeking high MPS jobs. Assignment of a low GNS person to a high MPS job cannot be expected to result in satisfactory productivity. Conversely, a high-GNS individual will be unchallenged and unproductive in a low MPS job. Figure 4.2 shows the four-cell matrix that characterizes those MPS/GNS match-ups and mismatches. We will explain the 4-cell match-ups, then will provide examples.

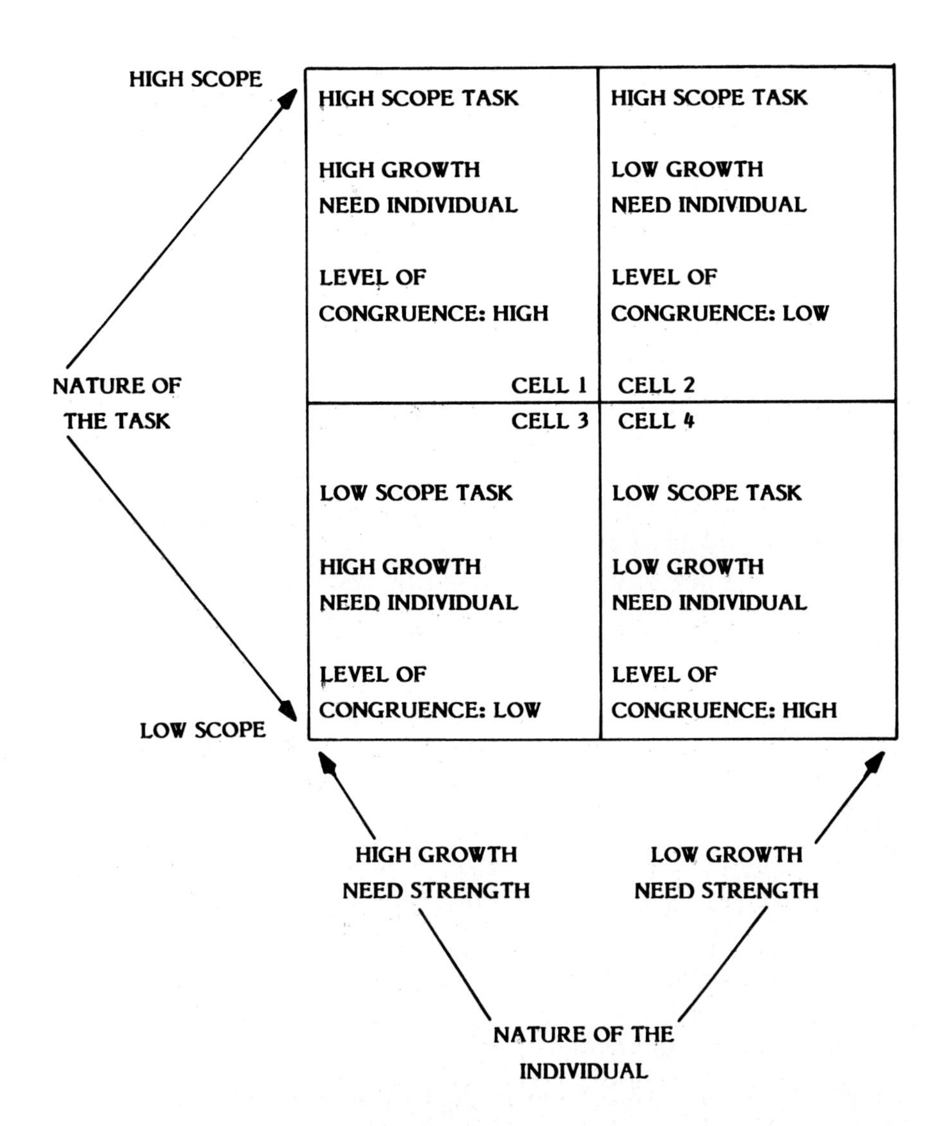

Figure 4.2: Level of Congruence Resulting from Degree of Match Between GNS and Task Scope

Cell 1—High Scope Task + High GNS = High Level Congruence

Individuals in Cell 1 have high-GNS. They desire tasks that are challenging: complex, nonstructured tasks which they perceive as significant. They want a high scope task for which they feel responsible and are aware of expected performance. If we have a match in scope of task and GNS, we have a high level of congruence between task and individual need. Examples of a high scope task are new applications development, data base design or telecommunications design.

Cell 2—High Scope Task + Low GNS = Low Level Congruence

Cell 2 reflects the situation where an employee with low GNS is assigned a high scope task. The level of congruence is low and a poor motivational environment results. An example is promotion of a person from operations, where GNS is lower, into the system development area. One would not expect good productivity in this situation unless changes are introduced to correct the disparity between task/individual need.

Cell 3—Low Scope Task + High GNS = Low Level Congruence

In the situation depicted in Cell 3, an employee with high GNS is assigned a low scope task. The level of congruence is low. The employee is likely to be frustrated and dissatisfied. There is no assurance that the level of productivity will be high. An example is routine maintenance activity asigned to a high GNS person.

Cell 4—Low Scope Task + Low GNS = High Level Congruence

The fourth cell reflects the situation where lower GNS employees (like the former computer operator described above) are assigned less demanding tasks, such as routine program maintenance. A high level of congruence exists between task and individual need. The motivational environment is good and a concomitant high level of productivity can be expected.

LACK OF GNS/MPS MATCHUP IN PRESENT MAINTENANCE ASSIGNMENTS

Two curves have been plotted from the research data (Figure 4.3) to show the mismatch of MPS and GNS for many of the assignments in the survey organizations. One would expect the GNS curve to parallel the MPS curve in order to utilize the congruence possibilities depicted in the four-cell matrix. Figure 4.3 shows that the parallel effect exists for the first two maintenance categories: 0 to 20% and 21 to 40% maintenance activity. Thereafter the curve sweeps up to form a U pattern with wide discrepancy (non-congruence) in the 41 to 60% and over 80% maintenance categories. This demonstrates how little understanding the field has of the GNS/MPS match-up required for good motivation. It shows that management has assigned maintenance to employees whose GNS is higher than the company norm--just the opposite of what they should have done. This is not an indictment of management; it reveals that lack of knowledge of motivation principles has intensified the problem. It also shows that GNS/MPS match-up is not an intuitive process—to the contrary, management's intuition has been counterproductive in this case.

Nevertheless, some individuals in the high-maintenance categories are quite productive. For these individuals, the GNS/MPS match-up has been achieved. Those circumstances will be examined next.

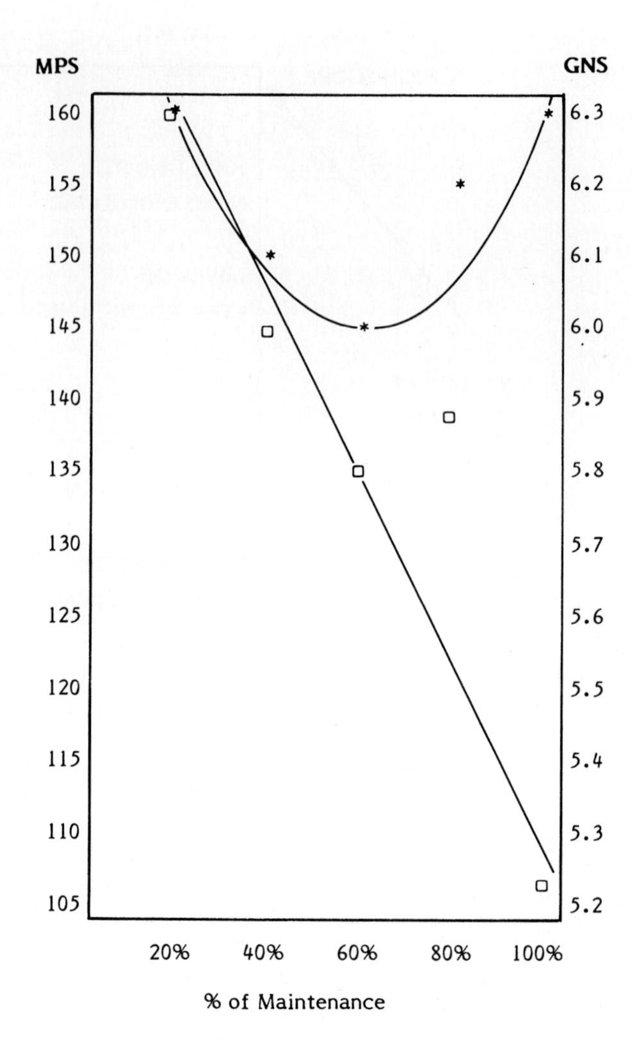

Figure 4.3: Lack of Congruence in MPS/GNS In the High Maintenance Categories

RESTRUCTURED TASKS FOR MOVE FROM CELL 3 TO CELL 1

With maintenance constituting approximately 50% of the workload, the ideal situation would be a DP department which had 50% of its personnel with low GNS and the other 50% with high GNS. New system development could be assigned to the high GNS personnel and maintenance to the others. The Cell 1/Cell 4 situation would exist—high congruence and high productivity could be expected.

Unfortunately, that ideal situation exists in few organizations.

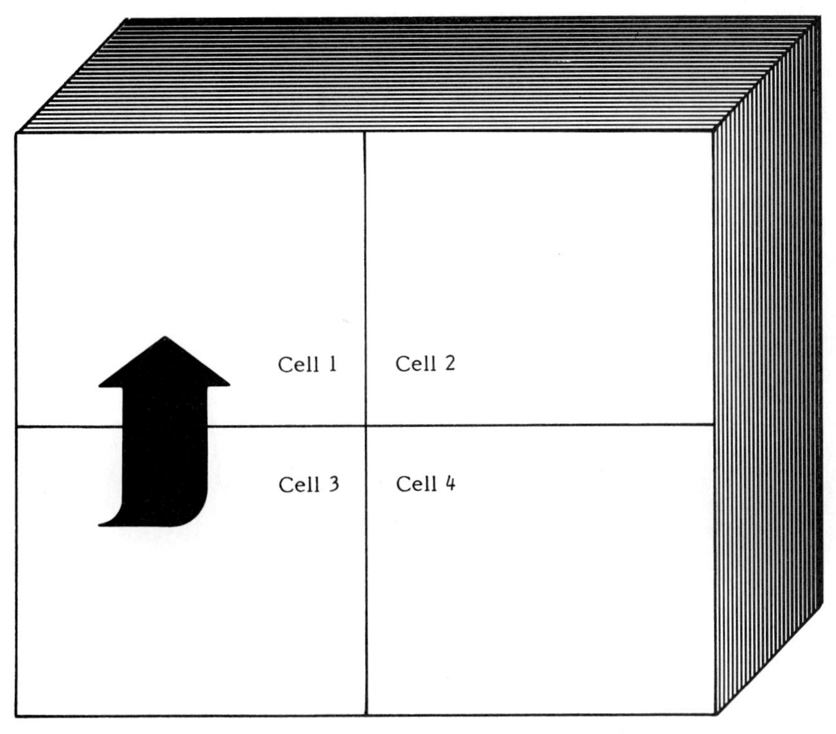

Figure 4.4: Objective of Restructuring Tasks to Move From Cell 3 To Cell 1 to Obtain MPS/GNS Match-Up

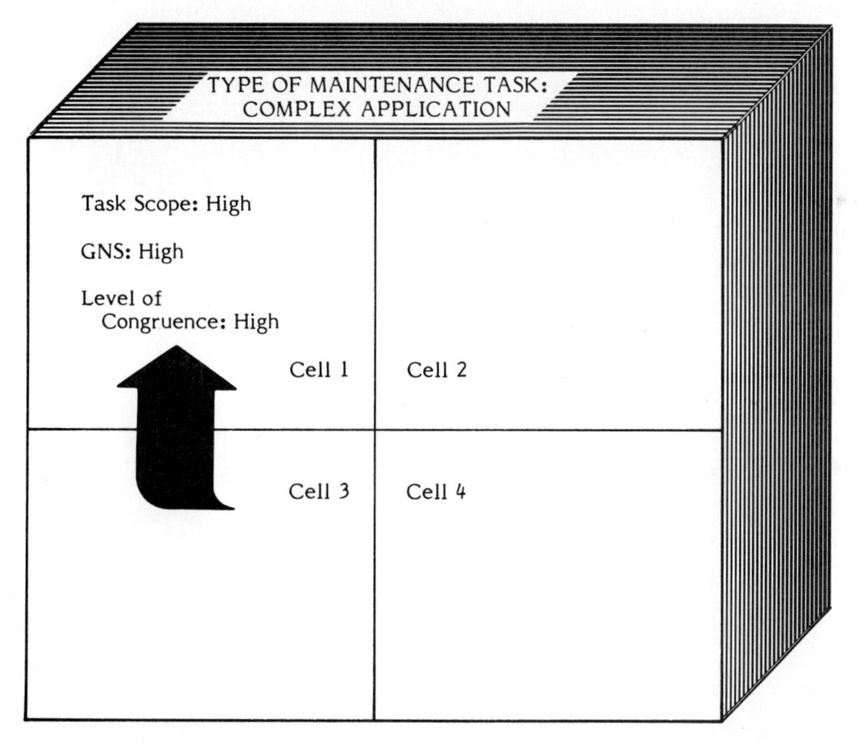

Figure 4.5: Added Dimension to Portray Job Characteristics Of Complex Maintenance Tasks

Some applications have a high degree of complexity - e.g. a management science application or a data base application with a large number of data interrelationships. Maintaining such a system can produce a high level of congruence for a high GNS individual.

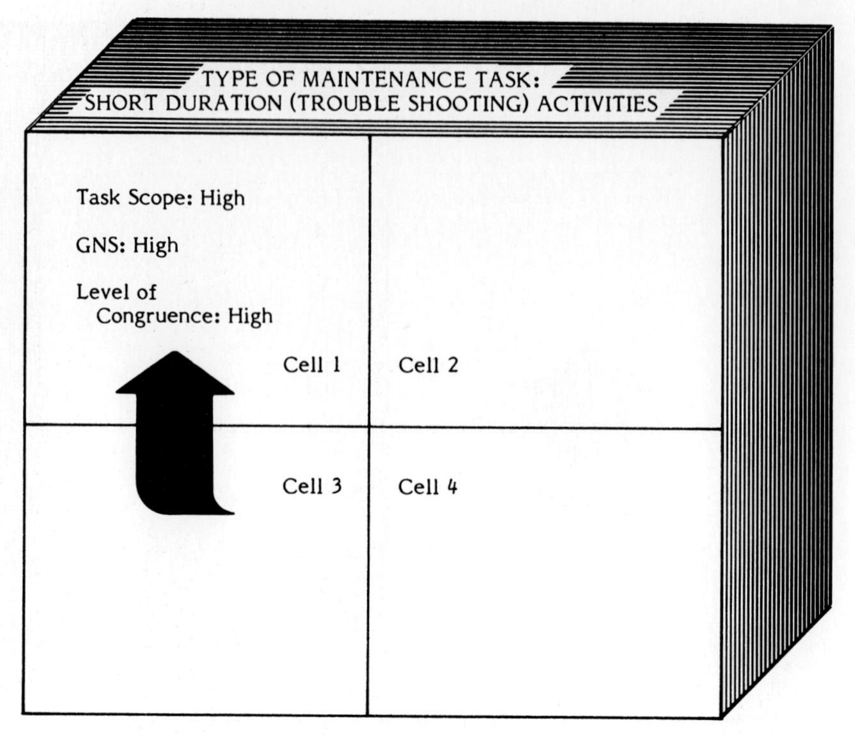

Figure 4.6: Added Dimension to Portray Job Characteristics of Short Duration (Trouble Shooting) Maintenance Tasks

Some individuals enjoy the challenge of "trouble shooting" - problem solving under critical schedule/critical application conditions. These persons are typically high GNS individuals. However, other high GNS persons would not enjoy these conditions. Typically, low GNS persons would prefer not to work under these conditions.

As discussed in Chapter 2, programmers and analysts have the highest growth need strength of any profession—higher than any of the 500 job types studied by Hackman and Oldham (see Table 2.2). The standard deviation around the mean GNS for analysts and programmers is also quite low, compared to other occupations. This means that the majority of computer professionals have high GNS. Therefore, the Cell 3 imbalance situation applies for the majority of analysts and programmers in companies where maintenance is over 50% of the work load.

Maintenance is perceived to be a low scope task by most computer professionals, therefore the less-productive Cell 3 situation would apply to a large portion of the organization's work.

What is needed is a way to restructure the work, to reallocate tasks, such that Cell 3 activities can be moved to Cell 1, as shown in Figure 4.4. That is, the work needs to be enhanced in scope.

Fortunately, the research revealed methods to restructure tasks to move work from a Cell 3 mismatch to a Cell 1 match-up. To illustrate the variety of approaches to accomplish this goal, the two-dimension, four-cell matrix needs a third dimension. That third dimension will represent either job characteristics or personnel characteristics.

Figure 4.5 portrays such a situation, where a maintenance task is challenging because of the level of complexity. Some high-GNS employees would find the work stimulating—resulting in a high level of congruence. An example is an application where complicated mathematical algorithms are utilized. Another example is maintenance of a complex operating system or a data base management system.

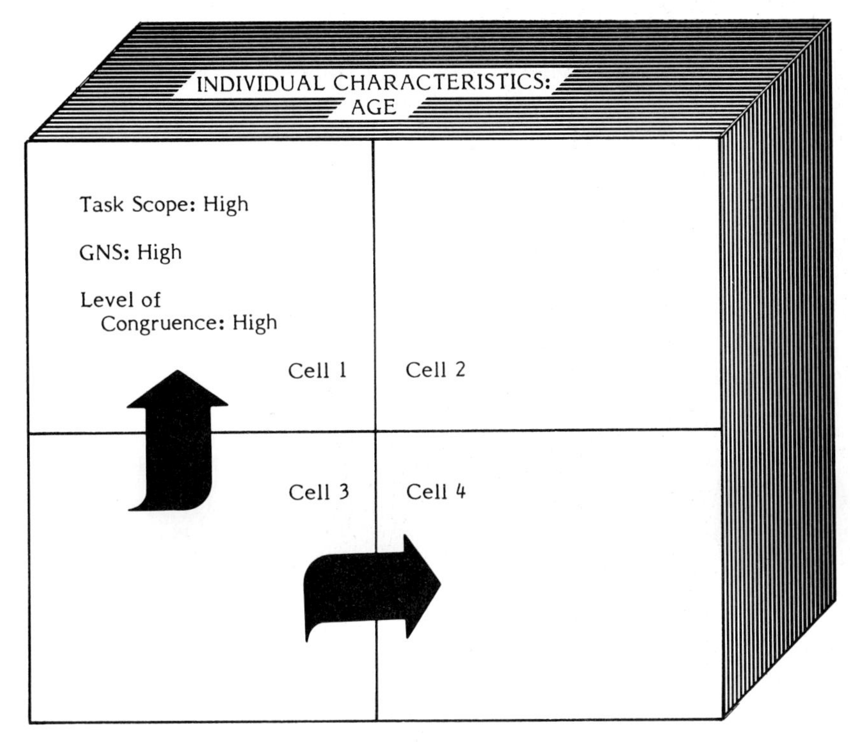

Figure 4.7: Added Dimension to Portray Individual Characteristics of Age

Some high-GNS, entry-level employees find a maintenance task challenging. This task is temporarily perceived as high scope and therefore produces congruence. After the employee learns the system, he/she perceives it as a low scope task. The other arrow depicts older employees, where GNS declines with age, as more likely to find congruence with a maintenance activity.

Figure 4.6 portrays a type of maintenance activity which is considered low-scope work by some employees and high-scope work by others. The "trouble-shooting" function on daily production runs is considered quite challenging by some high GNS personnel. The feedback is fast and the task significance is high because daily production runs have high visibility.

Other high-GNS people dislike short duration work. They are more challenged by large scope tasks. In other words, people who are motivated by short duration work are normally high-GNS persons. Paradoxically, other high-GNS personnel do not enjoy short duration work. Not all personnel know how they will respond to short duration activity and supervisors need to experiment in this regard.

Figure 4.7 portrays an individual characteristic (age) which affects level of congruence. Care in selection of tasks to assign individuals could produce congruence by moving a Cell 3 situation to either Cell 1 or Cell 4. The first example, movement of Cell 3 to Cell 1, occurs when a new, entry-level employee perceives a maintenance task as challenging. The new, high-GNS employee is productive in such a situation--until he/she becomes so familiar with the application that it is no longer perceived as a high scope task.

The other situation depicted in Figure 4.7 is the movement from Cell 3 to Cell 4, due to age. The Couger-Zawacki studies revealed that GNS was negatively correlated with age, as shown in Table 4.1 below. Therefore, older employees are not seeking scope of work at a level as high as younger employees. They are challenged by some maintenance assignments; congruence results, along with good productivity. It is interesting to note that the GNS of DP personnel over 50 is still higher than that of other professionals. Reference back to Table 2.2 shows that mean GNS for other professionals is 5.59.

It should be noted also that these are mean values. The standard deviation is 1.03. This means that some persons in the over-50 age category have a very high-GNS and need challenging work.

AGE	20-29	30-39	40-49	Over 50
GNS	6.089	5.835	5.776	5.635

Table 4.1: Effect of Age on GNS

INCREASING MPS BY ENHANCING SELECTIVE JOB DIMENSIONS

The research also identified other ways to enhance MPS when one of the core job dimensions is constrained. The survey results show that the skill variety job dimension is a key cause for employees to perceive maintenance work as less challenging. However, to lower MPS to the non-challenging level, more than one of the core job dimensions must be low. For example, the autonomy variable may be rated low by individuals—not necessarily because their bosses do not give them freedom to operate—but because procedures or policies provide little flexibility. An illustration is the constraint caused by procedures under which the original system was designed and programmed. Examples are: 1) the inability to use structured techniques to maintain a system because the original system was non-structured, and 2) having to write changes in an obsolete assembly language because the programs were originally coded in that language.

However, the key point to recognize concerning MPS is that it is comprised of five variables: skill variety, task identity, task significance, autonomy, and feedback from the job. Even though a supervisor may be unable to redesign the job to enhance all five dimensions, he/she can work on enhancing two or three of the dimensions to compensate for the low value of the other job dimensions. Stated another way, if skill variety and autonomy are low, perhaps task identity and task significance can be enhanced enough to raise MPS enough to produce the proper GNS/MPS match-up.

In several of the survey organizations, maintenance personnel were quite remote from the users of the system—little interaction occurred. In such a situation, the importance of the work (task significance) was not conveyed to the maintenance personnel. DP management could ask users to make presentations to the persons who maintain their systems—stressing the importance of their work. Some organizations physically locate maintenance personnel in the user area—to improve communication. Such an arrangement would also facilitate the enhancement of task significance.

Task identity could be enhanced in a similar manner. Lack of task identity would occur where an individual was working on a module of a system with little awareness of how that module relates to the whole system—or, how that system relates to the "company system." Supervisors

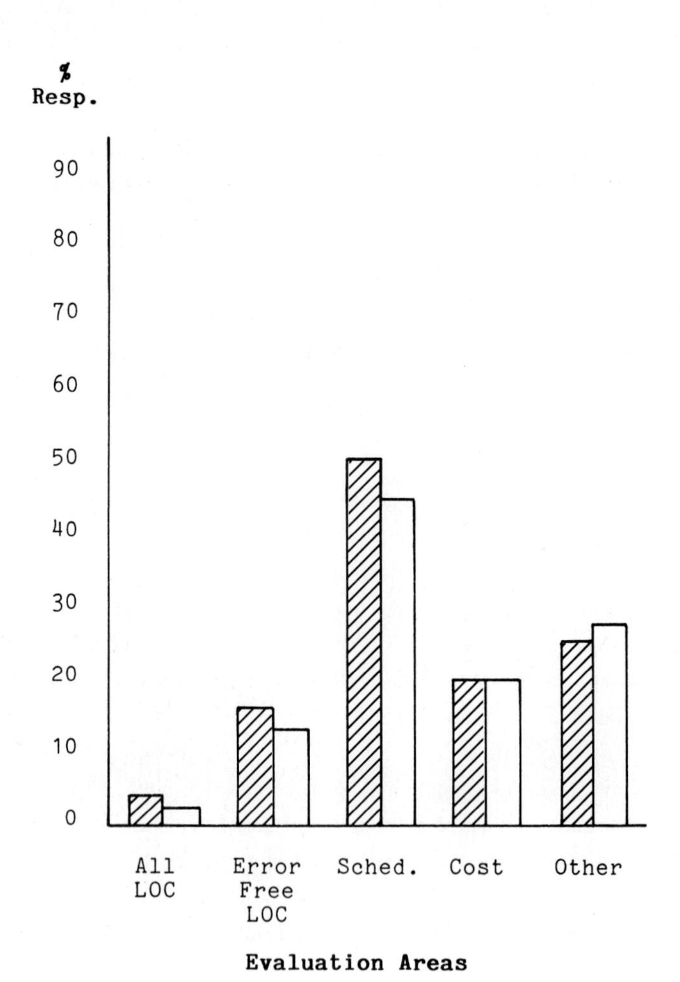

Figure 4.8: Perceived Areas of Evaluation of Maintenance Efforts By Interviewed Subjects

can place more emphasis upon showing these relationships to employees— thus enhancing task identity. The other component of task identity is completing a whole and identifiable piece of work, that is, doing the job from beginning to end. An example is working with the user to define the needed change, revising the program, testing and implementing the change. In some organizations, the programmer does only portions of that sequence of activities—for example, being told by an analyst specifically what lines of code to revise.

Closer interaction with users can also enhance the feedback dimension. However, improving feedback from the job means insuring that feedback mechanisms are designed into the work itself. Figure 4.8 identifies the performance evaluation criteria perceived by maintenance personnel as most important in their promotion.

Schedule compliance was by far the most important evaluation factor. It is also a factor that best illustrates feedback from the job. An employee does not have to be told by his/her supervisor whether the schedule has been met. Employees can determine that performance on their own. Companies that provide good project management (PM) systems—with emphasis that the information is primarily for employees and secondarily for supervisors—are enhancing feedback from the job. Some management goes a step further—to ensure that employees perceive the PM system foremost as their feedback system. Reports are provided to project team members several days before supervisors receive them. Problems are identified and corrective actions are determined by the employees and are presented to supervision at the same time they receive the PM system output.

The remaining job variable to be analyzed for enhancement, when skill variety is constrained, is autonomy. High-GNS personnel are highly goal oriented. For such persons, participative goal setting is important. They also seek freedom in selecting the manner in which

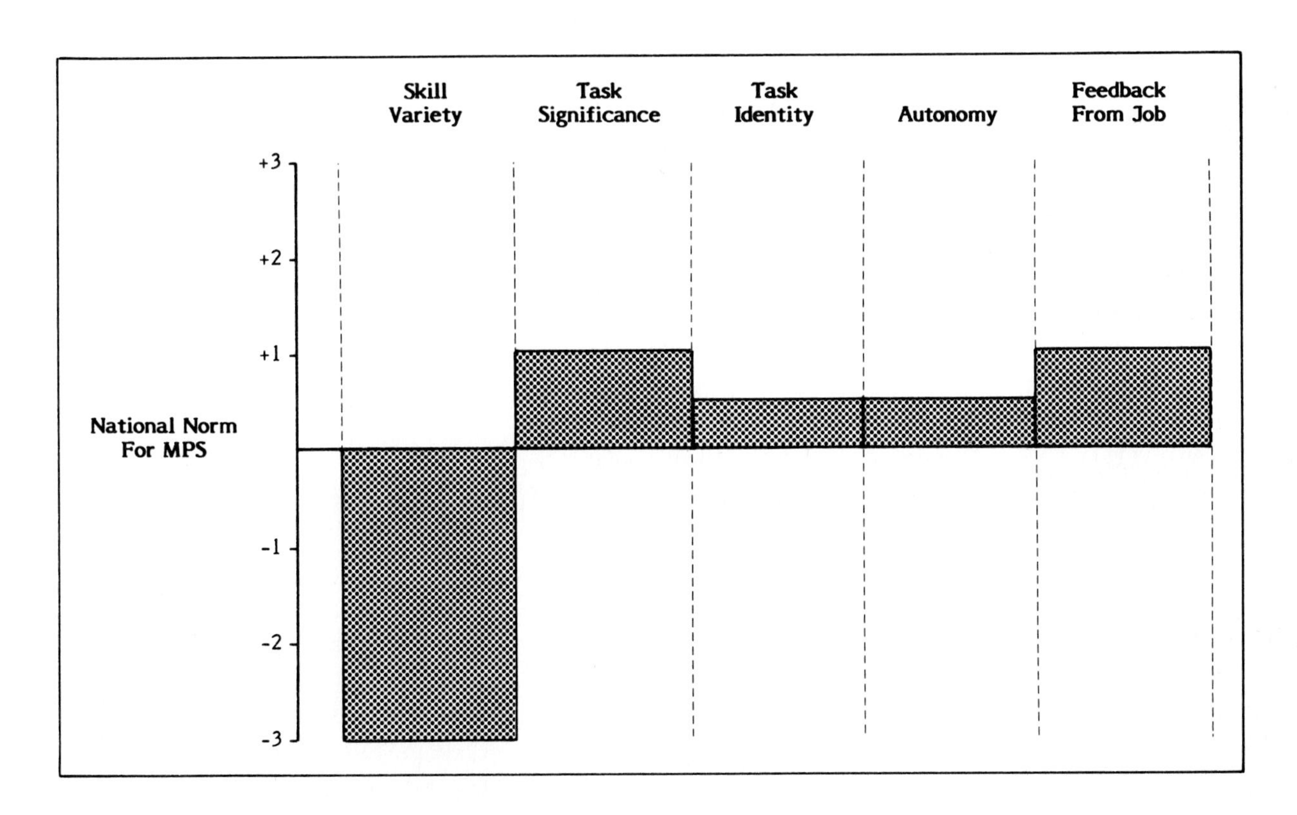

Figure 4.9: Restoring MPS to the Norm

Here one job dimension, skill variety, is severely constrained by the nature of the original system design. The other job dimensions have been enhanced to compensate for the low skill variety.

High GNS Characteristics

- Goal Oriented
- Ambitious
- Has Perspective (Separate Important From Less Important)
- Interested in Further Education/Training
- Self Starter
- Internally Motivated
- Confident
- Need for Recognition
- Assertive
- Inquisitive
- Systematic

Symptoms of High GNS

- Frequently Request Additional Training
- Seek Increasingly Difficult Assignments
- Seek Feedback—Want to Know How They Are Performing
- Want to Know Alternative Career Path
- Take the Initiative—Do Something That Needs to be Done Without Being Told
- Not "All Talk" About Wanting to Get Ahead But Takes Initiative—Self Study, Self Education
- Willing to Work Overtime (Unless Interferes With Self Education)

Table 4.2: Characteristics/Symptoms of High GNS Personnel

goals are attained. Supervisors of high-GNS personnel can enhance the autonomy job dimension by mutual goal setting, then refraining from close supervision during performance of the tasks agreed upon.

Even the skill variety dimension can be enhanced for maintenance employees. That job dimension is comprised of two elements—a variety of activities and use of a variety of skills and talents in carrying out the tasks. When the use of skills is constrained—such as confinement to a batch application when the person has the skill to work on on-line systems—the other component of skill variety can be emphasized. An illustration is assigning two people to jointly maintain two systems instead of requiring each to specialize on one system. Skill variety is enhanced for both employees.

Figure 4.9 summarizes the effect of these enhancements on the core job dimensions not constrained, where other core job dimensions are constrained by the characteristics of the system being maintained.

RECOMMENDATIONS

The research revealed rich possibilities for improving the productivity of maintenance. The C-Z Survey and interviews provided data to take these possibilities out of the speculative arena and to identify specific employee/job characteristics to improve motivation.

Employees assigned to maintenance tasks, who were productive in the view of their supervisors, were interviewed to identify the reasons. The preceding figures illustrate some new GNS/MPS match-up possibilities to enhance motivation.

Supervisors can work on enhancing the job dimensions not constrained in a maintenance assignment, as follows:

1. Skill Variety—assigning a variety of tasks where use of skills is constrained by the design of the system being maintained.
2. Task Identity—emphasis to better identify how the modules being maintained relate to the system as a whole and how that system relates to the company-set of systems. Also, being able to complete the whole maintenance task—from user interaction to producing workable code—will enhance task identity.
3. Task Significance—providing maintenance personnel an opportunity to work directly with users to recognize the importance of their work.
4. Autonomy—mutually setting goals with maintenance personnel, then allowing them to accomplish the work without close supervision.
5. Feedback From the Job—establish mechanisms to enable employees to track progress.

Determining MPS and GNS, in order to better match personnel to tasks, can be performed without the use of the C-Z Survey. Analysis of each job dimension can be accomplished, as described above, to compare MPS of various jobs. To identify GNS, supervisors can array their employees according to how well the attributes in Table 4.2 apply to each person (each attribute is a characteristic of a high GNS employee). The cases in the following section provide an in-depth analysis of approaches for enhancing the maintenance activity to improve motivation.

PLAN OF ACTION

We have found the following plan of action to be quite effective in improving the motivation of maintenance personnel:

1. Conduct a survey of employee perceptions concerning the motivational environment.

   The C-Z Survey instrument (See Appendix I) may be used for this purpose, although the organization may prefer to design its own instrument. It is important to preserve anonymity to insure candor in question responses. Individuals should be told that their responses will be tabulated by job type within the organizational unit and that individual responses will not be identified. For the same reason, it is also advisable to omit names from the survey answer sheet. The survey should include all personnel rather than a sample. This approach insures representativeness of the survey data. The survey takes only about 20 minutes to complete.
2. Process and organize survey results.

   Employee perceptions on the degree to which the five core job dimensions exist in their jobs are the principal survey objective. The survey results should be reported by job type within each organizational unit. Below that level, the mean response on each core job dimension is reported; that is, responses on the five core job dimensions/job/organizational unit. Record the results on the first page of the Structured Interview Questionnaire on Motivation (SIQM). See Appendix II for these forms.
3. Analyze results.

   When the employees in any job category indicate problems in two or more of the five core dimensions, improvement analysis (Steps 4-8, below) should be undertaken. However, our experience is that even those areas with satisfactory responses can be improved through the same procedure. Obviously, the problem areas are the prime candidates for improvement.
4. Determine causes.

   Part I of the SIQM is used to identify causes of these problems and targets for improvement. Each employee is provided the survey results for his/her job category (for that organizational unit only) and then asked to record his/her ideas on specific items to improve the situation.
5. Brainstorm Improvements.

   Armed with data from the survey and the SIQM, the unit supervisor schedules meetings of employees for each job type and leads a brainstorming session on improvements. The session is structured around the five core job dimensions. For example, the supervisor begins with a statement such as "Our programmer/analysts rated skill variety low compared to the national norms. What things do you think we might do to increase skill variety?" Employees bring their SIQM forms to the meeting to start the brainstorming activity.
6. Evaluate suggestions.

   The purpose of brainstorming is to generate as many ideas in as short a period of time as possible, e.g. one hour. The evaluation of those ideas is accomplished in a separate session. The ideas are categorized according to resolution constraints. That is, some suggestions can be implemented within the unit; others require approval beyond the unit. For example, reallocation of tasks is within the unit supervisor's purview. Approval at a higher level may be required to acquire a new software tool to facilitate program development. Changing to a new language may require approval of a company level standards committee. Even where a suggestion cannot be implemented due to organizational or financial constraints, open discussion of these factors brings about a better manager/subordinate relationship.

7. Prioritize suggestions.
   Priority of implementation should also be a group decision. Involving all affected employees attains greater commitment to the process. Some near-term suggestions should be given priority to give visibility to the project and to generate enthusiasm for implementing other improvements.
8. Assign project officers.
   To insure that these suggestions are implemented, the supervisor needs to assign project officers for each related set of suggestions.
9. Report results monthly.
   Realization of motivational improvement and, hence, productivity improvement, requires constant attention to implementation progress. A monthly progress report keeps all project officers on their toes and avoids de-motivation of other employees who might otherwise think their suggestions were being ignored.

CONCLUSION

Every organization is comprised of low and high GNS personnel. Use of the four-cell matrix can aid a supervisor in optimal assignment of personnel. Lower GNS individuals are more likely to perform satisfactorily on routine maintenance. Promoting operations personnel, who typically have lower GNS, into maintenance programming is another way to match GNS and MPS. Further refinement of individual GNS needs, such as desire to work on short duration versus long duration tasks, can result in an optimal assignment. However, most programming and analysis personnel have high GNS. Routine maintenance has low MPS and, therefore, produces a mis-match for these high GNS employees. The tasks should be enhanced to produce an MPS/GNS match. The Plan of Action is a proven procedure for enhancing maintenance activity to increase its MPS to that needed for a match with high GNS personnel. The cases in the following section provide an indepth analysis of approaches for enhancing the maintenance activity to improve motivation.

Throughout this report, we have referenced the Couger-Zawacki diagnostic survey. The implication may have been that you must conduct a C-Z Survey in order to realize the benefits cited in the report. That is not the case. We have provided data, analysis and observations to enable you to use this report in a standalone manner to improve motivation of maintenance personnel.

If you want to compare data on your own employees' perceptions to the data in this report, you need to consider conducting the C-Z Survey. A description is provided in Appendix I, along with the savings expected.

Likewise, it would be beneficial to have your whole management team trained in motivation theory and its application to computer personnel. Dr. Couger offers a 2-day in-house training course which provides such a background. Dr. Colter offers a three day technical seminar on improved techniques for performing maintenance on existing systems.

U.S. organizations are spending millions of dollars annually to maintain systems. Better congruence between scope of job and individual GNS will reduce turnover and increase productivity. With maintenance already consuming 50 percent of the labor budget of the average organization, and with the projection of even higher maintenance levels, improvement in this area is essential to ensure cost effective systems.

SECTION 3:
ACTUAL CASES OF MOTIVATION IMPROVEMENT FOR MAINTENANCE PERSONNEL

ACTUAL CASES OF MOTIVATION IMPROVEMENT FOR MAINTENANCE PERSONNEL

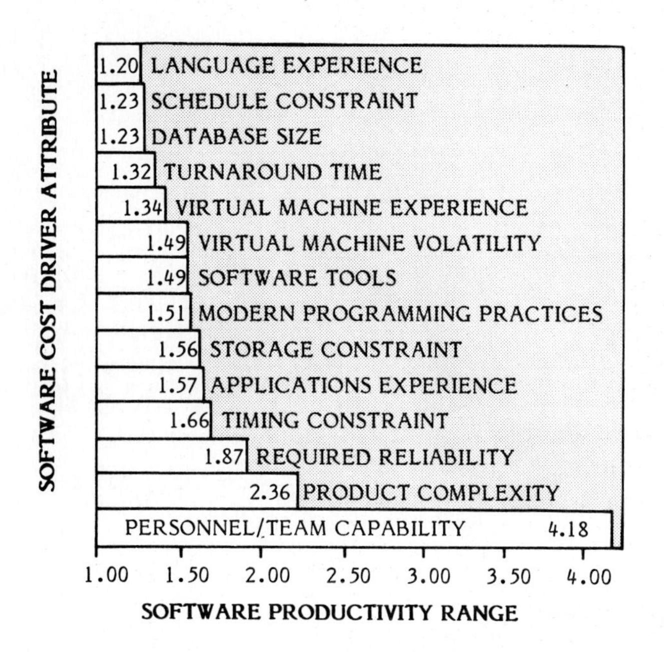

This illustration depicts the cost-drivers for software development, derived by Barry W. Boehm in his 767 page landmark publication, Software Engineering Economics (Prentice-Hall, Inc., 1981, p. 642).

Despite the improvement in the technical area, the most important cost-driver is personnel—by a wide margin. Improved approaches to motivating personnel are the key to cost-effective systems.

HIGH PRODUCTIVITY LOW COST
LOW PRODUCTIVITY HIGH COST
0.3 0.5 0.7 0.9 1.1 1.3 1.5 1.7 1.9
0.70 DEVELOPMENT TOOLS 1.42
1.02 0.98 DEVELOPMENT METHODS
0.90 1.11 SPECIFICATION VOLATILITY
0.67 USER COMPLEXITY 1.48
0.82 HARDWARE 1.22
0.81 GENERALITY 1.23
0.78 QUALITY 1.28
0.56 HUMAN FACTOR 1.80
0.88 1.13 CONTRACT

This illustration depicts productivity-drivers for software development, derived by the Nippon Electric Company (NEC), the highly regarded Japanese computer manufacturer. (Yukio Mizuno, "Software Quality Improvement", Computer, March, 1983, p. 69).

The figure shows that personnel factors not only have the more significant effect on software productivity but they also have the greatest negative effect. De-motivated personnel can have a huge adverse impact on a project's cost/schedule compliance.

CHAPTER 5
APPLICATION CASES

INTRODUCTION TO APPLICATION CASES

The firms selected for the research project have been quite creative in devising approaches to enhance motivation of maintenance personnel. Some enhancement activities were peculiar to that company's unique situation and are not capable of generalization. On the other hand, many enhancement activities are "portable" in the same sense that software can be generalized.

Those examples are recorded in this section of the report, using a standard reporting format:

1) Identification of the specific motivational issue or issues.
2) Explanation of the resolution process.
3) Evaluation of the results.

The names of the individuals and companies have been changed. Some of the problem areas were very sensitive. The purpose of this report is not to isolate poor managers and inadequate management practices but to identify effective motivational approaches. None of these cases are hypothetical. Neither are the cited improvements.

The generalized solutions can be applied in most companies with expectation of similar positive results.

Maintenance programmers can be motivated. Productivity can be significantly improved. The following cases substantiate these statements.

APPLICATION CASE A
TOP-DOWN APPROACH TO MOTIVATION ENHANCEMENT

An enhanced motivation environment is normally produced by either of two approaches: 1) the bottom-up feedback approach or 2) the top-down feedback approach. Each has advantages; each also has disadvantages. Pros and cons of each approach will be reviewed below:

Top-Down Feedback Approach

Using the top-down approach, managers analyze each job under their responsibility. Using the framework of the core job dimensions, the manager determines which dimensions are deficient. Use of the Couger/Zawacki survey instrument pinpoints deficiencies specifically. However, a manager who has carefully studied the motivation material in Chapter 2 of this report has an adequate understanding of the core job dimensions that can be enlarged by further study of Chapters 6 and 7 of the Couger-Zawacki book or by attendance of the Couger-Zawacki 2-day course on motivation of computer professionals.

A manager can perform the motivation analysis for his/her own area of responsibility independently of the other departments. This approach is usually undertaken by persons who—on their own—have attended the C-Z course or read the C-Z book. It is not easy to persuade the other managers who have not undergone a similar "unfreezing" process. The figure below illustrates that process.

Unfreezing occurs through 1) review of the data on computer personnel, gathered by use of the C-Z Survey, 2) analysis of the difference between computer personnel and personnel in other professions and 3) recognition of the potential of motivation job analysis in the computer community.

Change occurs when the manager accepts the value of motivational analysis and decides to implement this new approach into his/her area of responsibility.

Refreezing occurs when the manager utilizes the results of the study to make the necessary changes <u>and</u> continues to emphasize (freeze) those concepts in his/her management practice.

Examples of Motivation Analysis for a Specific Job

In one of the companies where the research was conducted, maintenance productivity was low, in the opinion of both the DP manager and the user. The work environment will be described first, then the motivation analysis will be reviewed.

<u>Description of the job</u>. The job was maintenance of the inventory system. A team of three persons was involved. Change requests (CR's) were initiated by both the user and the DP manager. User change requests tended to involve the effectiveness of the system: revised output

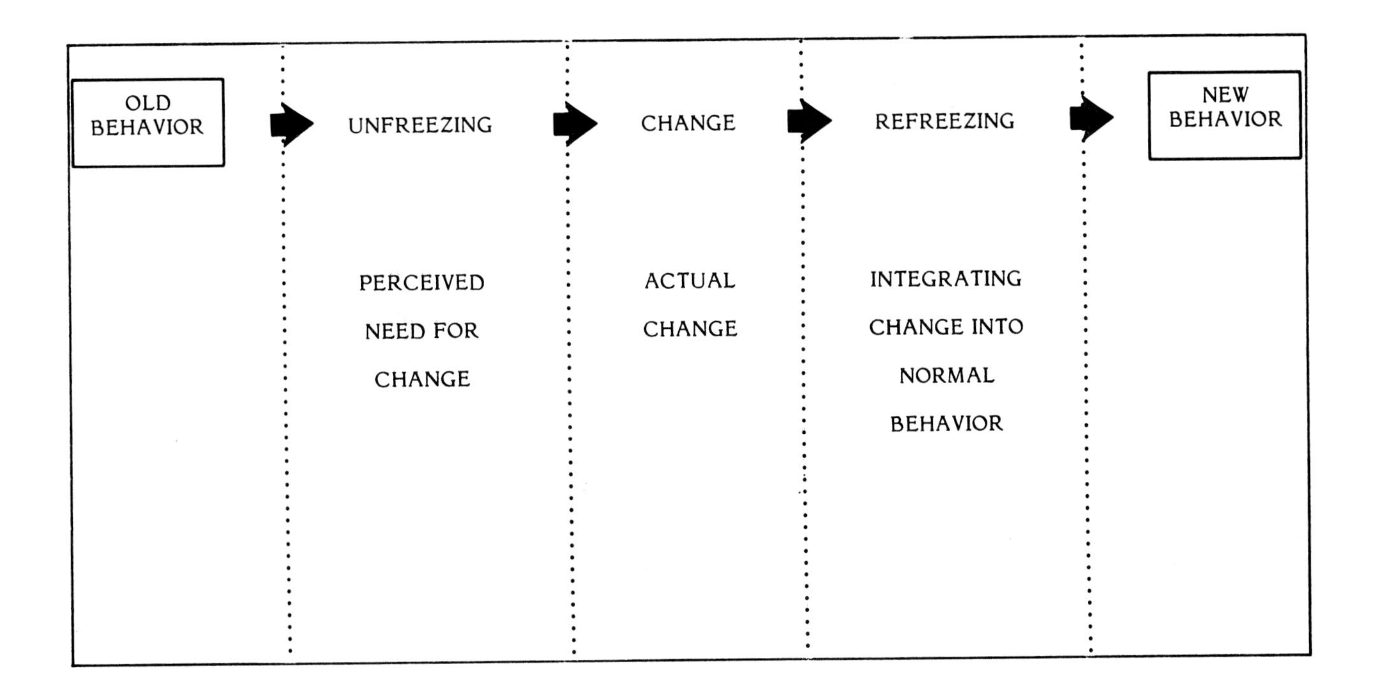

MODEL OF CHANGE

formats and incorporation of additional features. DP change requests tended to involve the efficiency of the system: speeding up through-put and sharing data with other applications such as cost accounting and purchasing. Both types of changes were negotiated by the supervisor of the maintenance group. An average of 20 lines of code was required to accomplish each change.

The system had been originally implemented five years previously and the documentation was weak, not unlike most DP applications. The code was unstructured; modules were quite large, involving several hundred instructions. In total, the system contained some 11,000 lines of code. The language was COBOL-68. The system had been patched hundreds of times by more than a dozen programmers who had progressed through this assignment to more senior positions. One of the three programmers presently assigned, Tom Schultz, had worked on the system for 2½ years—the others, 13 months and five months. However, the latter two people each had over 3 years experience in programming. Change requests were first analyzed by Schultz, who then suggested an approach to facilitate the modification. He also performed a quality review of the complete code and test results before they were passed on to the supervisor for approval.

Although input was on-line, processing was batch mode. The system was non-database oriented but shared data with many other files. Users were located in another building within walking distance. Over 30 people used reports from the system in their daily activities and another 15 or so were involved in inputting transactions to the system. The view of non-productivity of the programmers was based on their inability to meet schedule for completing change requests and to reduce the large backlog of change requests.

The programers were located in a cubicle near the elevator. They complained of the noise from people entering and exiting the elevator. They shared one terminal for programming and complained of not having individual terminals. One was a smoker and complained of having to leave the cubicle to smoke. The coffee dispenser and restrooms were at the other end of the floor (over 200 feet away) and they complained about the distance. They complained about their desks and chairs, which were almost ten years old while personnel on the floor above—the people working on the new on-line, database-oriented inventory system—had new furniture, two people to a cubicle with one terminal shared between them.

They complained about not being sent to the training course on the new computer where the inventory system was being developed and would be processed once it was completed. They complained about not being trained in the language of the new system, COBOL-74.

Finally, in response to performance reviews, they complained that schedules for change requests were often unrealistic—that management did not take into account the difficulty of modifying a poorly documented, heavily patched, non-structured system. They complained that test run turnaround time was poor because they were using the production computer instead of the computer devoted largely to development. The personnel working on the new inventory system were using the latter computer.

Motivation Analysis of the Problem Job

The manager of this activity, Norman White, also managed about 30 other people involved in maintenance activities. He had attended a Couger-Zawacki seminar on motivation of computer personnel. It was a public seminar and he was the only attendee from his company. Actually, he confided to the instructors, he selected this seminar because it coincided with his wedding anniversary and he had promised his wife to take some time off work and take her to a nice resort area.

Nevertheless, his high goal-orientation won out over his leisure time motivation, and he attended all sessions where previously he had planned to miss several sessions for additional beach time. He located an interesting city tour for his wife, instead. At the Thursday night cocktail party, his wife told one of the instructors that she had grown accustomed to his work having such high priority. "At least the time Norm spends with me and the children is high quality time—he works at making it so," she said. "So, I can't really complain. It's a lot better then having to deal with a husband who hates his job."

Back at the company, Norman was determined to apply his new knowledge. He chose as a test the problematic inventory maintenance job. He tutored the supervisor of that activity, Sarah Martinez, on motivation and asked her to read key portions of the C-Z book. He had already determined that Sarah's supervisory capability was not part of the problem, through discreet discussions with her employees. He and Sarah began a motivation analysis of the job. Sarah was a "quick-take," as are most high-GNS personnel, and quickly separated the hygenic factors from the core job factors (see Case G for further explanation of this difference).

"Most of those complaints are smoke screen," she commented. "It's their dissatisfaction with the assignment that causes these environmental factors to take on such importance to them."

"I agree," Norman said. "Let's concentrate on the factors directly related to their task." He used the framework of the core job dimensions as an analytical approach.

1. Skill Variety. They decided that skill variety was deficient. Work was confined to one application, using an old version of COBOL, and batch processing file orientation rather than on-line data base orientation.
2. Task Identity. Having the experienced person, Tom Schultz, conduct the initial analysis for a change request may save time for the short run, but it prevented the other two programmers from accomplishing a complete change request. The remote location of the user also made it more difficult for face-to-face discussion on the background of the change request. Their not being involved in the change request screening negotiations also reduced task identity for the programmers.
3. Task Significance. Lack of interaction with the user reduced awareness of the importance of their work. The new inventory system seemingly had greater importance because of the up-to-date hardware and software. Better facilities (new furniture, more terminals) reinforced that feeling of difference in significance.
4. Autonomy. Unilateral screening of change requests by the supervisor also reduced autonomy for the three programmers. Initial analysis by Tom Schultz reduced the feeling of autonomy for the other two programmers. Having their results reviewed by both Schultz and Martinez also reduced autonomy for these two. Autonomy was also lessened by having to work within the confinement of the existing program design and language. Lack of terminal

availability and computer test time lowered autonomy.

5. Feedback. The small size of the tasks should provide strong and frequent feedback—a perfect example of feedback from the job itself. If feedback from others—supervisors and users—is equally strong, this core job dimension is well provided for.

MPS/GNS Comparison

After completing the analysis of individual core job dimensions, Norman and Sarah felt confident that they had determined a major cause of the lack of productivity. The job was motivationally deficient.

They discussed at length the other factor which might have resulted in lack of schedule compliance—the validity of schedules. Sarah referenced the satisfactory performance of her other groups where she was the principal in schedule negotiation. "I believe I know the inventory system as well as I know these other systems and that the schedules for inventory changes are just as realistic as those for other groups which are meeting schedule."

Then Sarah and Norman discussed Growth Need Strength (GNS) of the inventory group. "When I compare characteristics of these three people to other people in the unit, I see them as average in growth need," she said. "I arrayed all my people, using the GNS characteristics chart you brought back from your seminar. These three are near the middle of the array."

"It looks like we have a mismatch in the individuals' need for growth (GNS) and the job's ability to provide that growth (MPS)," Norman responded. "As presently designed, this job is not rich enough to challenge these people."

Work Redesign

Having determined that there was a mismatch in GNS and MPS, Sarah and Norman began the process of redesigning the job to enhance motivation. They continued to use the analysis framework of the five core job dimensions.

Skill Variety. "Concerning the tools and techniques, we have little flexibility," Sarah said. "We could allow the use of structured programming techniques, but it make little sense to change the design when the system will soon be replaced by the new inventory system."

Norman agreed. "However, we can provide variety by letting them work on other applications. We could combine the teams maintaining the inventory and purchasing systems to increase skill variety for both groups. The reduction in specialization has some disadvantages, though, that might offset the benefits of increased variety. Let's work on the other job dimensions first and come back to this one only if we can't enhance the others enough to raise MPS significantly."

Task Identity. "We should let all three people review change requests, instead of Schultz's screening them first," Sarah said. "The disadvantage of the less experience of the other two is more than offset by enlarging task identity."

"What do you think about letting them in on the negotiation of change requests with the user," Norman asked. "That way, they'd be in on each phase of the change process and have their task identity increased significantly."

"If I did that for all my groups, it would be pandemonium," Sarah replied. "Why have a supervisor?"

"Could you at least let them sit in occasionally—to better understand the process and relate it to what they're doing? Remember, task identity is being able to identify or relate the tasks you're doing with the whole process."

"I agree," Sarah said. "We can do something similar at the tail end of the process—by letting the two newer programmers participate with Schultz in the quality review of the completed change."

"These changes will also increase skill variety," Norman added.

Task Significance. "Permitting the three to occasionally

sit in on the change request negotiations should increase task significance," Norman said. "Can you think of other ways to help them recognize that the work they're doing is important?"

Sarah hesitated, "That's tough. It's obvious to everyone that the new inventory system is the significant work. Although" —she added— "I might arrange with the project manager on that system to invite my people in on some of the user discussions on implementation to make them more aware of the value of inventory systems to the company."

Norman brightened. "That's how we'll attack the skill variety problem, too. We'll start training them to take over the maintenance of the new system. With its on-line, data base, structured methology and COBOL-78 characteristics, skill variety will be optimum. They won't be able to use these new skills for almost a year but they will be encouraged in that they are keeping up-to-date in their field. Skill variety will definitely occur through their learning these new techniques although not as well as if they were applying them themselves."

Autonomy. "Our earlier ideas will increase autonomy, too," Norman said, "by letting each of them work a change request from beginning to end, without Schultz's intervention."

"Yes, but we can't do much about the autonomy constraint of their being confined to the existing program design and language," Sarah said. "I'm afraid we're hamstrung on increasing autonomy otherwise."

Feedback. "The one job dimension that looks satisfactory is feedback from the job," Norman suggested. "The small tasks provide continuous feedback."

"However," Sarah interjected, "Because they're not meeting schedule, the positive feeling of completing the change request is greatly reduced."

"Could you let them be involved more in the process of schedule determination. That way it's their schedule too, not just yours," Norman asked.

Sarah frowned. "You keep coming back to that, don't you. I've got lots of CR's to negotiate for all my groups. I just don't have the time to include all the programmers in these discussions."

"But, this is your only problem group," Norman replied. "Once these job enhancements take effect, productivity should improve and these people should begin meeting schedule. Thereafter, mutual schedule setting should be less time consuming. Why don't you give it a try?"

Sarah was reluctant. "O.K. I'll try it. I like the suggestions we came up with. It's worth the additional effort if I can get this problem resolved. I must admit that this group's lack of performance is affecting my own motivation."

Results

The problem was diminished. These job enhancements produced positive results both in output and attitude. The complaints about peripheral issues subsided.

Conclusions

The figure below depicts the sequence of activities in the top-down approach to motivation enhancement related to the change general process documented earlier.

However, we need also to examine the change process as related to the individuals involved.

The decision to attend a motivation seminar or to read a book on the subject indicates openness—a person willing to change behavior if the reasons appear valid. However, we used Norman White's example for another reason. His primary reason for attending the seminar was a few days of relaxation. That is not unusual in our field. Likewise, non-supervisory employees are often sent off to a course or seminar as a reward for good performance.

In Norman's case, he got hooked on the subject and ended up attending all the sessions. Although his initial decision to attend was conditioned by his need for a vacation, it is obvious that he had a real desire to improve motivation of his staff. This evidences his openness to new ideas and to change. The seminar discussions, citing specific data about computer personnel motivation, were an unfreezing

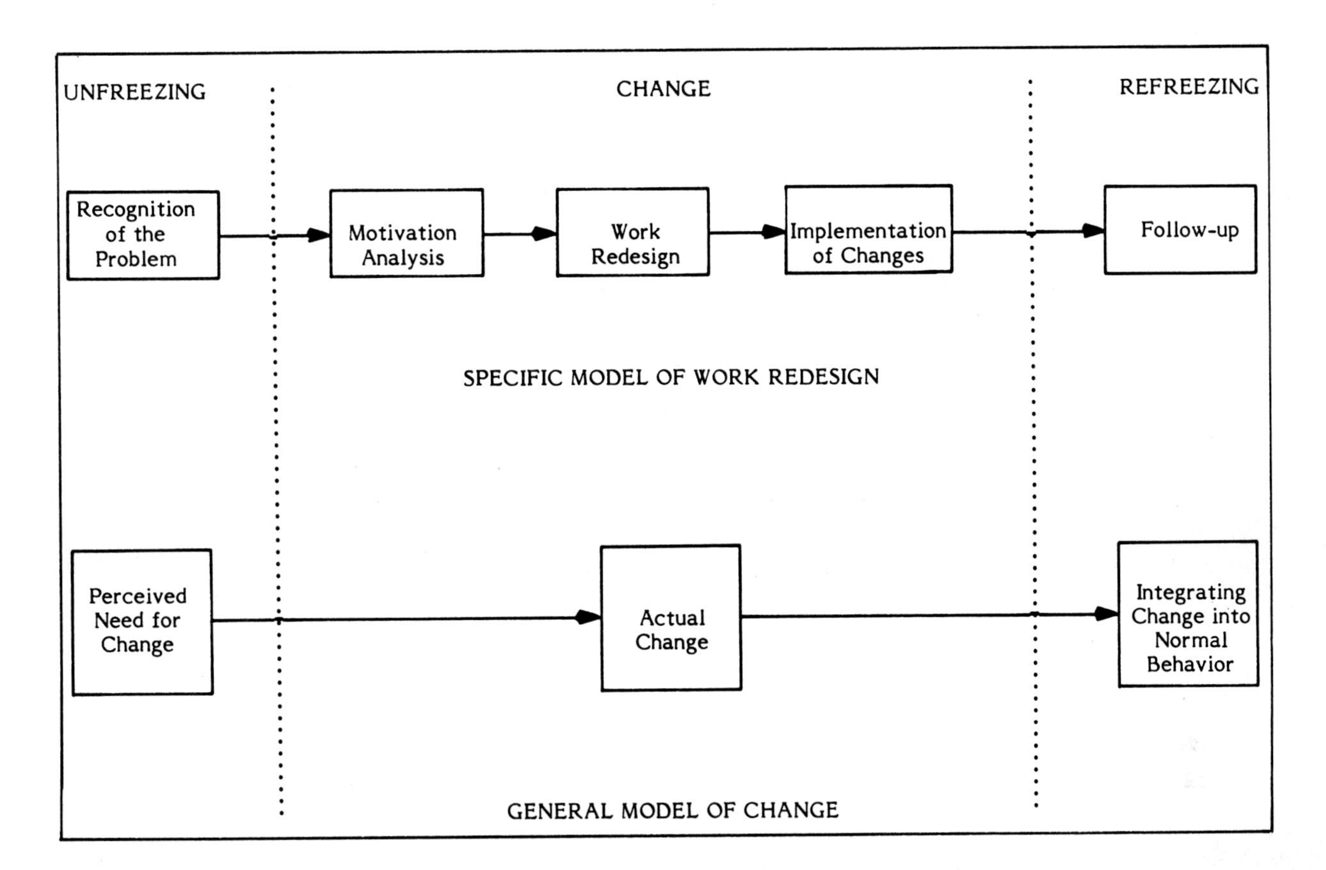

process for Norman. He went back to the job convinced that the approach would improve motivation of his employees.

Sarah's unfreezing process began with an external impetus—a request by her boss that she read the motivation material in preparation for a discussion of its application to her area of responsibility. She could have resisted the next step lower—the actual change. Yet she perceived the need for change through a combination of reading the suggested materials and recognition that she had not been able to solve the inventory group problem through other means.

Fortunately, the employees also "bought into" the change process. They saw the benefits and became committed to implement the changes. If all three—the manager, the supervisor and the maintenance personnel—continue to utilize the motivation enhancements agreed upon—to refreeze themselves into the new behavior—the results will continue to be positive.

Not all employees are willing to commit themselves to a change process in which they are not involved from inception. For them, the bottom-up approach, discussed next, is the better approach.

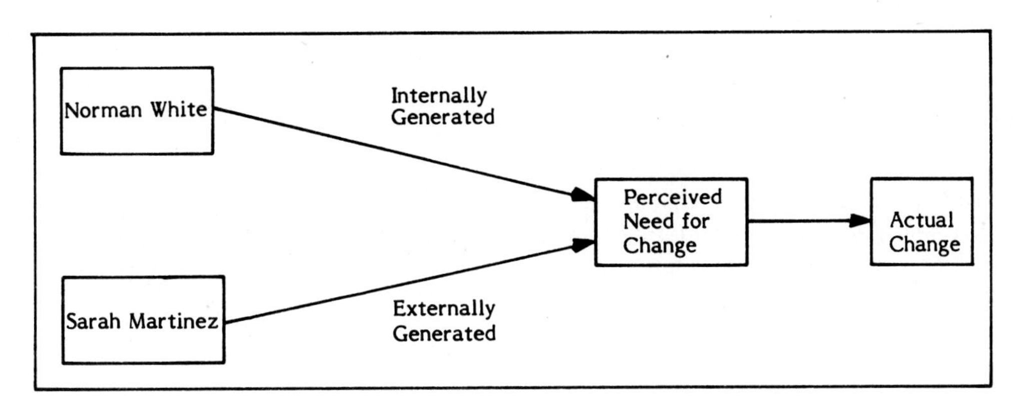

APPLICATION CASE B
BOTTOM-UP APPROACH TO MOTIVATION ENHANCEMENT

The author once consulted for a firm where a profit-sharing plan was introduced with unexpected negative employee response. It was a generous plan, where an employee who spent 25 years or more with the company could realize a quarter-of-a-million dollars or more in personal income upon retirement. Nor would there be any reduction in benefits from the existing retirement plan.

Management was sure that the company's high turnover rate would be reduced significantly as a result of the new program. Conversely, turnover actually increased the next year due to employee rejection of the way the plan was developed and constituted.

The principal problem concerned access to benefits. To begin with, employees could not enroll in the plan until they had been with the company for two years. Second, to realize maximum benefits, the employee had to contribute up to 10 percent of his/her salary into the program each year. This amount, along with an equal company contribution, would be invested for the employee as a part of the program. Third, a severe penalty was imposed for early withdrawal. Employees could withdraw their own contribution, but would forgo all company contribution. Further, the new program was announced on Christmas eve, implying that it was a gift of the owners rather than a recognition of the contribution of the employees to the company's profitable position in the industry.

Had employee representatives been involved in the planning of the program, they would have conveyed these important needs that management overlooked. Instead, a formal, grievance-like report (the company was non-union) was provided to the owner anonymously. It was nailed to his door, like Luther's 98 theses concerning the Catholic church were nailed to the cathedral door. In angry words the report chastised the owner for ignoring the needs of the employee. It explained that they did not need a big lump payment at retirement. It went on to explain that there were two periods in an employee's career where normal salary was insufficient to meet the typical employee's financial need: 1) when a home is purchased and 2) for tuition for college age children. The typical employee would have to save 10 percent of his/her salary for these two important needs, thus would not be able to participate in the program to the extent expected by management. They would, therefore, realize much less than the advertised benefits of the new program.

Management chose not to revise the program and a number of employees retaliated by leaving the firm.

Benefits of Employee Participation

The behavioral science literature is replete with examples of the superior results of employee participation in the design of a program, rather than merely participating in its implementation. The same logic applies to a motivation enhancement project. The bottom-up approach insures that employees' views will be identified and lowers the probability of their resistance to change.

The Bottom-Up Procedure

While the bottom-up procedure is also appropriate for introducing change in one unit, it is especially effective for organization-wide changes.

For a motivation enhancement project to produce an optimum result, all managers should be involved. An in-house seminar of two-days duration will provide the knowledge acquisition necessary for a successful motivation project. The seminar provides an unfreezing process. Then a survey is undertaken to determine employee perception on motivation.

The other approach is for the company to conduct the survey for all employees, then gather all managers to discuss the results. The survey results are the unfreezing process; they cause managers to want to acquire knowledge about motivation enhancement.

Either approach is effective. The term, bottom-up procedure, pertains to the way in which the survey results are disseminated. It consists of the following steps:

1. Feedback of survey results to employees by 1st line supervisors.
2. Brainstorming ways to resolve problems identified by the survey (after tutoring employees on motivation concepts).
3. Evaluation and prioritizing of ideas.
4. Report to the next level management on survey results and suggested changes.
5. Report to a steering committee set up to handle recommended changes which impact more than one unit. (Committee is comprised of both managers and non-managerial employees.) For this category of suggestions the committee tabulates, evaluates and prioritizes suggestions. It also prepares a report to management.
6. Discussion, evaluation and prioritization of suggestions by management.
7. Explanation to employees on reasons for priorities.
8. Appointment of project officers to implement changes.
9. Preparation of newsletter to keep employees informed on progress of implementing suggestions.
10. Monitoring of progress by steering committee.

Dual Commitment

A two-way or dual commitment occurs in use of the bottom-up approach. First, when management asks employees to complete the survey, they in effect are saying, "We are committing to careful evaluation of what you are telling us in the survey results and will attempt to implement the changes you tell us are needed. However, for reasons of company policy or resource limitations we may not be able to implement all your suggestions.

Approach to Motivation Analysis

	Bottom-up		Top-Down
Advantages			
1.	Employees have a say, thus more committed to change.	1.	Less time required to identify improvements.
2.	Larger number of improvements identified.	2.	Management has more control of the process. Steering committee not required.
Disadvantages			
1.	Top managers only see report on persons reporting directly to them.	1.	No assurance that employees committed to implementing changes.
2.	Top managers must wait until lower level units complete their analysis and pass it up.	2.	Total effect may be sub-optimal.

However, we guarantee an explanation on every suggestion."

When employees participate in the feedback sessions, they are more apt to commit to positively working out acceptable solutions. The fact that they are involved in the feedback sessions and in identification of improvements usually assures employee commitment to the process.

Comparison of Two Approaches

Each approach to motivation analysis—top-down or bottom-up—has its advantages and disadvantages, identified on the preceding page.

The following figure conveys the greater complexity associated with the top-down approach. However, it cannot convey the higher level of commitment—and thus—higher probability of success of the bottom-up approach.

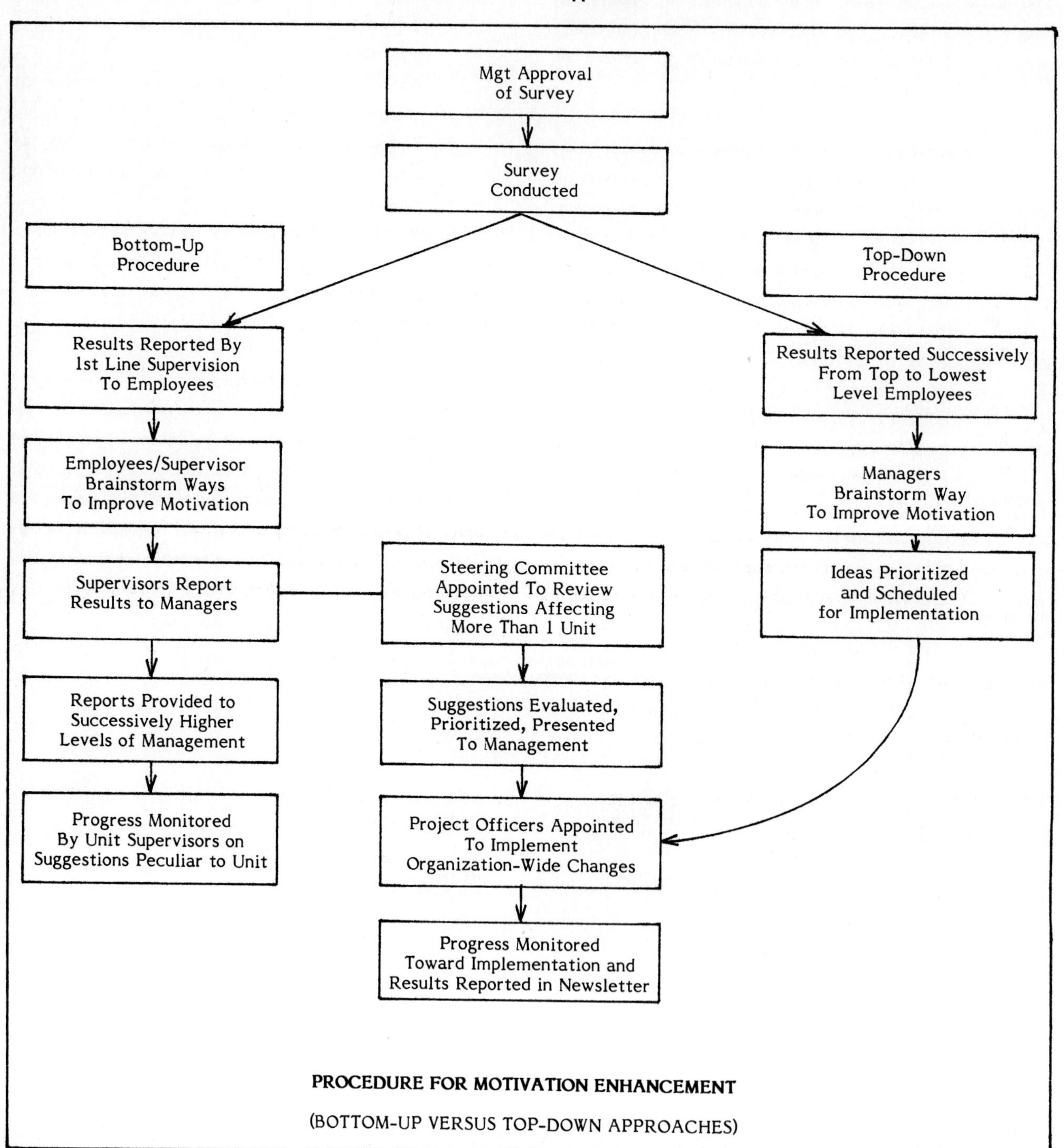

PROCEDURE FOR MOTIVATION ENHANCEMENT

(BOTTOM-UP VERSUS TOP-DOWN APPROACHES)

APPLICATION CASE C
MATCHING GNS AND MPS TO ASSURE PRODUCTIVITY

Most of the cases thus far have concentrated on only one of the two components in the formula for productivity—the work itself. The cases demonstrated how the core job elements were analyzed and enhanced to raise the job's MPS (motivating potential score).

This case will concentrate on the other component in the productivity equation—GNS. The strength of an employee's need for growth determines the degree of richness needed in task assignments. If the two are balanced, motivation will be optimized.

Deficit in GNS

We will begin with the description of a company situation where a major imbalance in MPS/GNS occurred. The C-Z Survey revealed a company mean for GNS well below the national norm. In fact, it was the lowest of any company we've ever measured. This is a FORTUNE 500 company with up-to-date computing equipment and a good budget.

Because of inadequate performance of the DP organization, the previous vice president was replaced with a person of proven management ability from another area of the corporation. His knowledge of DP was minimal, however. He asked us to conduct the C-Z Survey to provide him some data about his motivational environment to determine if that might be the cause of the organization's inability to meet schedule and budget objectives.

The low GNS for all development units was a surprise both to us and to him. It had resulted in a significant mismatch in GNS/MPS. The company was heavily moving into on-line, database technology and the work was quite rich. The programmers and analysts were overstretched— their low GNS was not compatible with the job's MPS. Poor motivation resulted. Because the GNS/MPS mismatch was organization-wide, it was obviously the major factor in the performance deficiency.

After some effort, the cause of the problem was determined. It was the personnel department. That department performed the initial screening of all potential additions to the DP organization. When questioned about screening criteria, the screeners responded, "This company places great emphasis on harmony and cooperation. Therefore, we try to screen out all aggressive candidates."

Systematically, for a number of years, they had screened out high GNS personnel. Table 4.2 identifies the characteristics/symptoms of high GNS personnel. In that table, "assertive" is a synonym for "aggressive," although a more positive characterization.

This problem was not resolved easily. The screening criteria were changed immediately. However, it would take a long time for the future hires to raise the organization's mean GNS. Two approaches were undertaken:

1. Narrowing the scope of the job to produce a Cell

4, high congruence match-up, from the previous Cell 2, low congruence situation (see Figure 4.2, for the 4-cell comparisons).

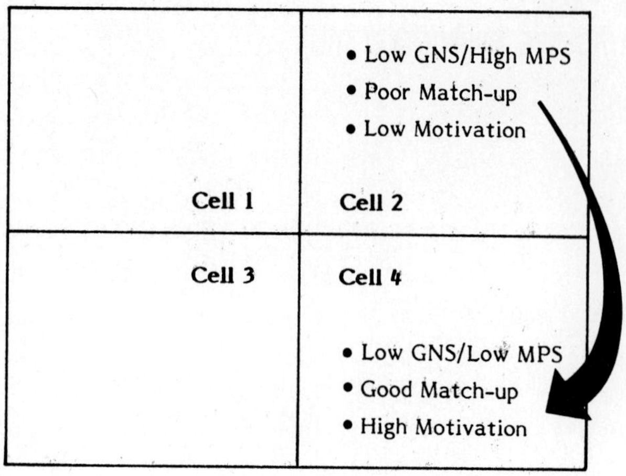

Cell 1	• Low GNS/High MPS • Poor Match-up • Low Motivation Cell 2
Cell 3	Cell 4 • Low GNS/Low MPS • Good Match-up • High Motivation

2. Assigning newly hired, high-GNS personnel to those activities where the job richness was hard to revise, that is, where the scope was not easily narrowed. Examples are telecommunication and database design. Here a Cell 1 situation occurred. A high-scope (rich MPS) job is matched against a high GNS individual.

With these changes, the company was able to enhance its motivational environment significantly—resulting in increased productivity.

Assessing GNS

How does a supervisor assess GNS of his/her employees? By use of Table 4.2, a supervisor can analyze the past behavior of each employee to determine the degree of conformance to those high-GNS characteristics.

We could have just as easily established the table to identify low-GNS characteristics. An example is the table of symptoms of low-GNS behavior at the right.

Trying to Emulate High GNS

We've been asked a number of times why we didn't publish these lists in the earlier book, Motivating and Managing Computer Personnel.

Some experience with students in our Information Systems degree program precipitated that decision. For example, one of our few easy-going students, whom I'll call Joe Modicum, keeps pestering us for this list. He's very candid about his reasons, "I want to be able to convince the recruiters that I am a high-GNS person."

You see, he is not a high-GNS person and wants to know the attributes of such a person so he can try to emulate them in an interview. While he doesn't have the acumen of our high-GNS students, he should at least recognize that all he has to do is observe those students and emulate their characteristics!

Fortunately, this student is not representative of the I.S. student body. I.S. majors typically exhibit the same characteristics as their industrial counterparts. They have selected themselves out of the general student body and are not threatened by the continuous re-tooling process ahead of them—to the contrary, they are excited about it.

A few students are attracted to the field because of the higher pay and more rapid chance for advancement. Joe Modicum is one of these people. An example of his low

Symptoms of Low GNS

1. Doesn't badger supervisor on career path
2. Easy-going
3. Rarely requests training but not adverse to it
4. More cautious about alternative career paths
5. Not as willing to work overtime
6. Rarely takes initiative; usually needs to be asked
7. More easily distracted from work list to other, less important tasks
8. Not as goal-oriented
9. Reactive vs. proactive
10. Has trouble making a choice or decision

GNS is his part-time work at Safeway all during his school years. It is not unusual for our students to have part-time jobs. Before they obtain some DP qualifications in their junior year, they take whatever jobs are available. Most I.S. majors quickly change to part-time programming jobs as soon as they've completed the two required courses on that subject. A few others, like Joe, continue to work in the job in which they've become comfortable. Also, they may be earning more by that time than they could as a beginning, part-time programmer. The high GNS students are more interested in the opportunity to get into their career field and will sacrifice salary to meet this goal. Joe bought a new car his senior year, instead!

We're changing our previous policy and publishing the GNS characteristics in this report because it will have limited distribution and will cost enough to be prohibitively expensive to students. It's inevitable, though, that you'll be interviewing people in the future who will somehow obtain the GNS characteristics list and who will be trying to convince you that they are high GNS people.

Should All Your People Have High GNS?

We may have given the impression that you should attempt to acquire only high GNS personnel. That was not our intent.

First, note that intelligence is not a characteristic listed in Table 4.2. You may have a very bright but unambitious employee whom you need for some high-intellect type work.

Second, not all assignments can be made equally challenging. Lower GNS personnel can be assigned the less challenging work, producing a Cell 4 match-up. Motivation will be good and, concomitantly, so will productivity.

While this situation is appropriate to the development area, it especially applies to operations. We were surprised to find GNS so high for operations. The national norm is second only to that of analysts and programmers—higher than any of the 500 occupations studied by Hackman and Oldham. Unfortunately, the MPS is not equivalently high. The MPS for operations is lower than any of the 500 occupations studied by Hackman and Oldham. A terrible mismatch occurs in the typical operations shop.

It is possible to enrich operations jobs, using the same procedures cited earlier. But even with enrichment, most do not need a person with GNS almost as high as the programming/analysis GNS. Generally, operations managers overhire—that is, select personnel whose GNS is higher than needed.

Some companies purposely hire operations personnel with higher qualifications than needed—good logical ability and high GNS—with the intention of moving them through operations into programming. Management in these companies must recognize, however, that the mismatch for these people in operations may cause low productivity during their tenure there.

Promoting Operations Personnel into Maintenance

Some of the firms in our research project have a <u>standard</u> promotion path from operations to programming. So long as the approach described above is utilized, the results should be satisfactory—hiring personnel whose qualifications match programming tasks and temporarily routing them through operations. These people will even be productive during their operations tenure if they clearly "see the light at the end of the tunnel." If they know they must perform well in this preliminary assignment in order to meet their longer term objectives, most high-GNS personnel can set short-term objectives that coincide with operations. On the other hand, if management is not able to fulfill its promise of transfer, productivity will decline. An example is a recessionary period, where growth is slowed and fewer openings occur in the programming departments. Even this situation is not insurmountable.

If it is a case of a promotion path for the typical operations personnel—those without programming qualifications—the result will be unsatisfactory. Such a career path should be selective, not standard. Using this approach, only those operations personnel whose qualifications match the programming prerequisites are selected for the alternative (non-operations) career path.

With that background, let us relate this situation back to

the topic of this case—assessing GNS to enable a match with MPS.

Although the average operations GNS is higher than any of the 500 occupations measured by Hackman/Oldham, it is significantly lower than that of programming personnel. Assigning these persons to development of new systems would produce a Cell 2 situation—low congruence.

However, assigning these persons to a lower scope task, such as the traditional maintenance activity, produces a Cell 4 situation—high congruence. The typical maintenance task, without enrichment, has an MPS that would match these lower GNS persons.

So, a selective promotion path from operations to maintenance programming is a viable approach.

Conclusion

By use of Table 4.2 (characteristics/symptoms of high GNS employees) and the table in this case (symptoms of low GNS employees) it is not difficult for a supervisor to array his/her employees according to GNS.

Our experience shows that high GNS and low GNS employees are quite easy to identify. The mid-range group is not as easy to array. The need to clearly identify where each of these mid-range GNS employees is ranked in the array is not essential, however. The fact that they are mid-range is the important determination. All employees in this group can be assigned rich jobs, relative to those with low GNS. The high GNS personnel, whose characteristics fit Table 4.2 very closely, should be assigned the highest scope tasks.

We're frequently asked why we do not provide a GNS Determination Test for managers to use to ascertain GNS for each employee. There are two reasons:

1. As employees get familiar with the test and begin discussing it, they find ways to beat it. As an example, one of our I.S. majors went on a recruiting trip during Spring vacation and visited 8 companies in 5 days. Six companies administered the IBM programmer aptitude test during that period. By the final time he had taken the test, he was so familiar with it that he had raised his score to the 99 percentile.
2. Our experience shows that supervisors are reasonably accurate in separating GNS into three categories: high, medium, low. Such classification is precise enough to achieve a GNS/MPS match.

APPLICATION CASE D
CHANGING THE NEGATIVE IMAGE ABOUT MAINTENANCE

An energetic and successful manager, Bill Hamilton, was assigned to direct the units responsible for maintenance. He knew that personnel in the DP organization generally considered maintenance as less challenging and interesting work, compared to new systems development. His boss, who reported to the Director of Data Processing, asked him to take on the new assignment "because you've been able to achieve good productivity from your people in your present assignment and we need desperately to improve the productivity of the maintenance activity."

He went on to say, "We're concurrently improving the homogeneity of tasks by pooling all the maintenance work and assigning it to you." He concluded, "We've discussed this move at length in the managers' staff meetings and agreed that your management record proves your ability to handle tough assignments—this will be one of your toughest!"

Bill had little to say about who was transferred to his new unit (the people most knowledgeable about the application were transferred along with the application). He inherited the typical mix of applications. At one end of the continuum were some applications written in assembly language for 2nd generation computers, being emulated on the few 3rd generation systems still in operation at the company. At the other end of the continuum were recently completed applications, written in high level languages and running on 4th generation systems.

The backgrounds of personnel varied just as much. Some were in their fifties—former users who transferred into the DP department when the functions they were working on were converted to the computer. They had been trained by vendors; few were college graduates. Others were recent hires—college graduates who had been thoroughly trained in the company's six month programming curriculum.

Hamilton's Goals in Improving Productivity

Bill took his new assignment seriously. He assumed that he would have the function for a couple of years and was determined to motivate his people to achieve the productivity levels expected by his management.

He decided on three principal strategies:

1) to demonstrate the importance of the work—that it was just as valuable to the company as new development projects
2) to better define productivity goals in maintenance and to track performance carefully
3) to award performance and highlight high achievers as an incentive to other personnel

Clarifying the Causes for the Negative Image of Maintenance

Bill called all his employees together and explained his goals. He concluded with the statement, "I recognize that some employees in DP consider maintenance as less interesting work. Would you share with me some of the things you've heard so we can analyze the basis for these views?" Note the positive approach in his invitation to discuss the issue. That approach had four key elements:

1) He did not try to cover up or circumvent the issue. He wanted it aired, with each sub-issue clearly identified so corrective action could begin.
2) He did not intimidate the group by stating disagreement with the view that maintenance was less challenging than new development work. He showed his interest in objective discussion which might lead to a change in attitude.
3) He did not polarize the group by asking some of them to state why they felt maintenance was unchallenging. He asked them what they had "heard" from others. Thus he created a non-threatening environment for discussion.
4) He invited them to use their analytical skills to examine the problem.

Using this strategy he decreased the likelihood that the session would become a bitch session. He revealed a desire to surface the real issues. He began the process of team building through encouraging cooperative resolution of the problem.

He turned a problem area into a challenge where people could use their natural analytical and problem-solving skills.

Categorizing the Complaints

The discussion went on for more than an hour. Bill recorded the comments on a flip-chart. When all the different views had been identified, he separated the people into groups of three and asked them to categorize the problems. Only a half-hour was required and the four resulting categories were represented by the following comments:

* "Some people look down on those of us performing maintenance—as if we aren't capable of the creativity necessary for developing new systems."
* "It's laborious and time-consuming to try to locate a bug in poorly documented software—especially when its coded in assembly language."
* "Patching programs is not nearly as challenging as writing your own programs from scratch."
* "A person who spends his time working on older systems is not keeping up with the technology—he'll quickly become obsolete in a field which changes so rapidly."

Identifying Ways to Improve Maintenance

Bill used the same process for identifying ways to change the negative image of maintenance work—he asked his people to brainstorm ways to enhance the maintenance activity.

Two benefits accrued from this process:

1) The creative talents of the whole group were utilized, resulting in a variety of ideas to choose from.
2) A commitment to the goal of resolving the problem was attained, through involvement of all the people.

Over 30 suggestions were obtained. A task force was then set up to evaluate and prioritize the suggestions. The group consisted both of supervisors and non-supervisors. This same group identified implementation approaches. A schedule for implementation, with associated budget, was presented to Bill Hamilton and, after some negotiated changes, was adopted.

Only one category of problems will be discussed at this point because we have discussed resolution of the remaining three categories in other cases. Category 1—concerning the importance of maintenance work compared to new development activities—will be discussed here. The other three categories are covered elsewhere, as follows:

Category 2 - problem of working with poorly documented programs. This issue is discussed in Case M.

Category 3 - problem of patching someone else's program instead of writing your own program from scratch. This issue is discussed in Case G.

Category 4 - problem of obsolescence due to work on older systems. This issue is discussed in Case H.

Changing Perceptions About Maintenance

"Is maintenance actually less significant work?" Bill Hamilton asked himself. After thinking for several minutes, he concluded, "No it isn't. These systems are important to the company. But why try to defend this position? We should be asking that question of the people for whom we're maintaining these systems."

Bill attacked the problem on two levels:

1) By obtaining statements from key users concerning the importance of their applications to company objectives.
2) By selecting effective ways to disseminate that information.

The first task proved the easier of the two. Users were quite willing to provide written statements about the impact of their systems. "Somehow," Bill reflected, "I must do more than just distribute these statements. It's a passive rather than an active approach and may not accomplish the change in attitudes that I'm seeking." He came up with several ideas but then decided that it would be better to let the task force work this problem. "Better to let the people closest to the problem—and the ones with most at stake—brainstorm solutions."

As expected, the task force produced several excellent suggestions:

1) Post the statements on a special bulletin board, with the caption, "Our Systems Are Vital to This Company's Principal Objectives."

2) Arrange for tours of the user area for all employees in the unit—to acquaint them with the user functions and how their systems were integrated into these functions.

3) Invite key user managers—ones in important jobs—to periodically speak to the unit, explaining how the reports are used and their significance.

4) Invite the DP Director to explain to the unit the importance of maintenance work compared to new development work.

Creativity Necessary In Maintenance?

The four suggestions were implemented and effectively began to change perceptions about maintenance being less important than new development. However, these measures did not directly deal with the perception that maintenance activities "require less creativity"—implying that these tasks were assigned to personnel who had less creative ability.

Bill Hamilton believed this issue was just as important as the one concerning the importance of the systems being maintained.

"Certainly, some changes are routinely accomplished," Bill concluded, "However, some development work is likewise routine. A system designer specifies precisely what the system outputs are to be and the programming standards manual specifies precisely what programming methodology is to be followed."

Bill also identified other maintenance work that is far from routine—requiring special analytical skills. "Identifying the cause of a problem is greatly expedited by a person with good analytical ability. Likewise, locating the section of code to be changed in a poorly documented program is not something you assign to mediocre personnel. Special analytical skills compartmentalize the problem and lead to early resolution."

In his next unit meeting, Bill passed on this reasoning and asked his people to respond.

"I agree," one senior member commented. "It gives me a great deal of satisfaction knowing I'm one of only two people who understand this system and who can quickly identify a problem to get the job up and running again in the shortest time."

Another remarked, "The same thing applies to incorporating a change-request. You aren't going to be able to hire just anybody to accomplish these changes and do them efficiently as well as effectively."

That comment triggered one from a programmer who took special delight in being able to improve a system. "Regardless what you call it—special creative or analytical ability—it is the ability to dig into a system implemented by the development people and then improve it—to make it run faster or to improve the response time for a user sitting at a terminal. It takes a special ability to accomplish such an improvement."

Evaluation of Results

The end result of the motivational program introduced by Bill Hamilton was a cohesive team which prided itself in not only keeping key company systems running but also constantly improving them.

In his annual performance evaluation, Bill's supervisor confirmed that these accomplishments were apparent to the entire organization. He gave Bill a substantial salary increase.

The process used by Bill is not unique. It is described in Appendix IV (pp. 147-148) of the Couger-Zawacki book, under the heading "Procedure for Work Redesign." It has been utilized in dozens of companies with similar positive results.

The process elicits ideas on all kinds of motivational improvements, some hygienic (environmental) and some which are job-specific. In other cases, we will concentrate on the separation of those two motivation categories and identification of the relative importance of suggestions for enhancing motivation.

The category 1 problem in this case affected two important core job dimensions: feedback and task significance. The enhancement of the job dimension, task significance, is quite clear from the above description. The enhancement of the feedback job dimension is less obvious. Nevertheless, the activities described above increase feedback significantly. The closer interaction with users naturally provided better feedback. Some was formal through written statements of the value of the work. Other was informal through direct contact with the users.

So—the implemented suggestions directly affected two of the five core job dimensions. The improvement in motivation and productivity was a natural outcome. However, this improvement process is not confined to maintenance programming, of course. It is the process used to redesign and enhance any DP job. By involving those persons directly affected by the problem—those who understand it best—the probability of resolution is greatly increased.

APPLICATION CASE E
RAPE AND PILLAGE OF THE MAINTENANCE TEAM

As described in Case D, Bill Hamilton was assigned the management of the maintenance activity at the time it was consolidated. Previously each development group was responsible for maintenance of all applications in the user area it supported.

Hamilton used proven approaches to build a sound motivational environment and, by the end of his first year in the new job, productivity of his group was recognized throughout the DP organization.

In addition to concentrating on enhancing the core job dimensions of task significance and feedback, Hamilton worked hard at team building. By using proven behavioral approaches he succeeded in building a cohesive group with high esprit de corps.

Career Ladder Delineated

In addition to the factors listed above, Hamilton was interested in building a solid career path for his personnel. He negotiated with DP management and the Personnel Department to establish a career ladder by which maintenance personnel could progress to the senior level. The only higher position was that of technical specialist, the one position where a non-supervisor could earn as much as project leaders.

The section was comprised of 32 persons, including three supervisors. There were three administrative personnel reporting directly to Hamilton, so each unit averaged nine persons. There were three positions (programmer trainee, programmer, programmer/analyst). The latter two positions had three levels, i.e., programmer I, II and III. There were two paysteps in the trainee category and three for each level thereafter.

Matching GNS and Job Scope

Hamilton and his subordinate supervisors worked diligently at identifying growth need strength (GNS) of all employees in the maintenance section.

Job scope was carefully matched to the individual, utilizing the principles discussed in Chapter 4 of this report. For "fast burners" (high-GNS personnel), rapid progression up the career path was assured. Lower-GNS personnel progressed more slowly, but still in accordance with their need for growth.

There were a few employees who professed a need for moving ahead but whose performance was not consistent with their ambition. As explained in Case C, there are means to separate verbal declarations from actual behavior to ascertain the real levels of GNS.

Hamilton's supervisory team became proficient in GNS assessment during the first year in the new section.

Spotlighting the Maintenance Function

Hamilton's careful attention to sound management/motivation practices paid off in an unexpected way. His boss, Steve Arden, frequently cited Hamilton's accomplishments in the manager's staff meetings. In doing this, Arden was demonstrating the use of core job theory of motivation. Through this recognition Hamilton was attaining enhanced feedback and task significance.

Hamilton was quick to respond that most of his first year accomplishments were due primarily to a motivated group of employees. He was also quick to share the praise with his subordinates, insuring a constant flow of feedback for them as well.

The unexpected result of spotlighting his productive personnel was the attraction of his peers to these people when job openings occurred in the development area.

"It shouldn't have been a surprise, I suppose," Hamilton told his subordinates in one of their weekly staff meetings. "I think we should respond positively to these requests. It broadens the career paths of our people."

After a few months, however, Hamilton's immediate subordinates began to complain. "The development managers are using us as their training group," one said. "Instead of growing their own, they take one of our seasoned people."

"It's worse than that," another, Jane Claybaugh, said. "They are avoiding the time-consuming process of personnel selection. Why seek outside, when we'll let them have a proven product?"

The third supervisor, George Osborne, was equally adamant. "Because their turnover is higher than ours—due to our hard work at producing an exceptionally good motivating environment—they have more openings. There aren't as many vacancies in our section, so our people are inclined to transfer in order to move up quicker than we can promote them."

"We've carved out a virgin area that's now desirable to all others. It's **rape** and **pillage** of the maintenance section," Osborne charged.

Hamilton thought the characterization a little overdone, but he had to agree in principle, "We won't be able to maintain our level of productivity if we are overburdened with the task of training personnel for the development sections."

Osborne responded, "Won't our people begin to consider this section some sort of housing bureau? They're just with us until we can find a good home for them."

Hamilton was inclined to agree. "However," he said, "if we really believe in the career path philosophy, we won't stand in the way of a promotion for any one of our employees. Once they see us doing that, we'll lose credibility."

Claybaugh concurred, "Now I understand for the first time the expression 'on the horns of a dilemma'."

Resolution

"I think we can move off the horns," Hamilton responded. "We can at least tilt the horns in our favor. Can we agree that we don't want to prevent growth opportunity for our people? Meeting growth need is a prime motivator. We've accomplished it by expanding the scope of work and responsibility—enhancing the core job dimensions. One way is to promote a person to a wider-scope job. Another is to continue to expand the present job for that individual. However, we must reward performance or the employee will become demotivated, regardless of the scope of the job. So long as we can promote people within the section and keep them doing interesting work, we shouldn't have a guilty conscience about retaining them. In other words, we can discourage lateral moves. On the other hand, if one of our people has a chance to progress faster by transferring, we should encourage it."

He concluded, "We need to counsel our employees, however. These promotions are valid and a step up so far as the organization ladder is concerned. But—is the new job really as challenging? We will insist on evaluating it in terms of the core job dimensions so we can honestly appraise each new job opportunity for our people. There are definitely some jobs in the development area that are not as rich as our redesigned maintenance jobs. When we are assured the transfer is good in all respects, we will encourage the move. If not, we'll put the burden on the development sections to prove benefits of the transfer."

Evaluation of Process

Hamilton's logic was sound. It also precipitated action in the development sections to perform a motivation analysis of their jobs. His peers in the development section had a new incentive to introduce the motivation study like the one Hamilton had conducted (explained in Case D).

His credibility among maintenance staff was not only maintained, it was amplified. Although the experience level diminished somewhat, the goal congruity between employees and management spurred internal motivation and productivity.

APPLICATION CASE F
THE LEPER COLONY

One organization in northeastern United States has resorted to a unique approach to try to improve maintenance productivity.

Management in this company, which we will refer to as ELIC (Eastern Life Insurance Company), decided that the maintenance function should be separated from the rest of DP—not just organizationally, but physically. They believed that the attitude that maintenance was less interesting and less challenging was transmitted—like a communicable disease. "Let's try training a group of maintenance programmers from scratch," they decided, "isolating them from present personnel who have been instilled (infected) with the belief that maintenance work should be avoided." They agreed also to begin with people who have not graduated from computer science or business information system degree programs, where they may also have been told that maintenance work was less challenging. "If we recruit people from other disciplines," they concluded, "they'll be excited about the opportunity to get into the computing field, where salaries are higher and advancement is faster. They should be motivated and productive."

To insure the efficacy of their approach, they not only housed these personnel in a building across town from the parent company DP department, but also set up a subsidiary company. Both groups shared the same computer, located at the parent company.

Control on Positive Attitude Toward Maintenance

The only contact with the parent company was through senior analysts who carefully and formally specified program modifications for enhancement and who reviewed and approved Change Requests for "fix-it" maintenance activities. The selection process for these senior analysts included attitude questionnaires and interviews to insure assignment of persons who would support the new positive attitude about maintenance. All came from the parent company to maximize knowledge of the functions for which applications were being maintained.

Supervision for the new subsidiary was selected in a similar fashion—persons with company experience and positive attitude toward maintenance.

First Year Results—Positive

After the first year of operation (18 months from approval of the concept), management of the subsidiary met with the parent company DP management to assess results. All agreed that the results were positive. Only those personnel who performed well in the six-month probationary period were retained. They continued to perform satisfactorily. The communication problems of conveying programming specifications had been a significant problem during the first six months of operation, but after a year of operation there was general agreement that this process was an unforeseen benefit of the new approach—program specifications were better documented.

In addition to this qualitative assessment of performance, a quantitative system was instituted. The two principal components were schedule compliance for enhancements and response time for fix-its.

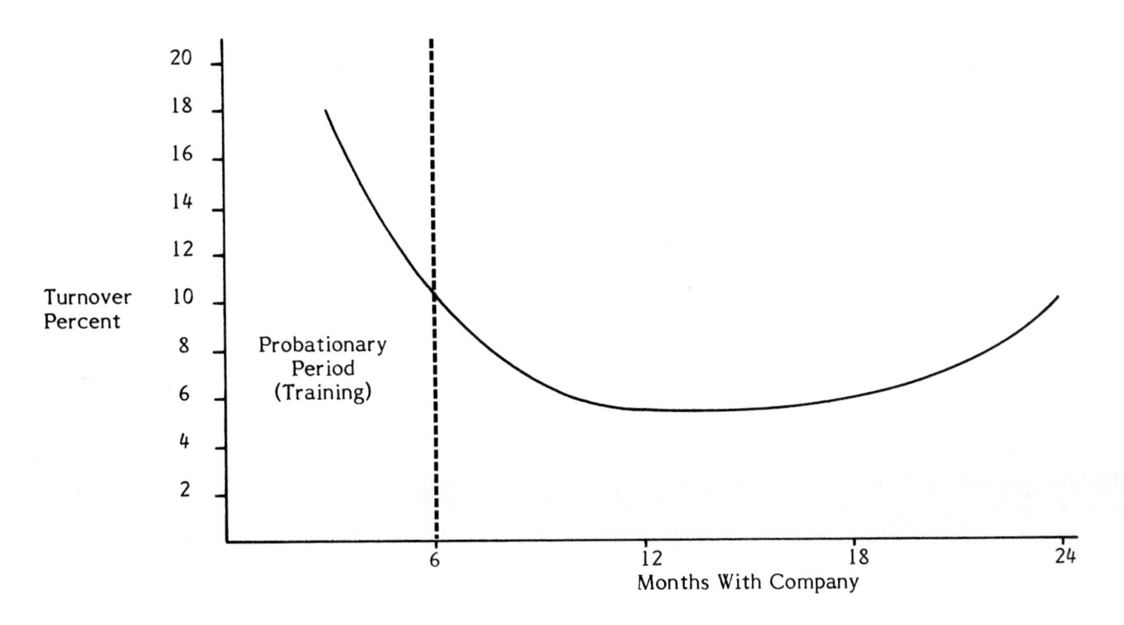

Second Year Results—Nominal

However, the second year performance assessment revealed several problems. Productivity declined in the fourth quarter. In addition, the subsidiary was beginning to experience turnover at the same rate as the parent company.

Several causes were identified: 1) with 1½ year's experience, personnel were becoming marketable, 2) career paths were more limited than those in the parent company, where personnel could also move into user management or operations management. (See figure above.)

One manager listed another cause, "Some of our people are joining professional societies and are hearing negative comments about maintenance from fellow members." His colleagues agreed. "You can control information exchange within the organization, but you can't control the external channels," one responded. "You can't even control your own employees off-site," another said. "One of our maintenance people met a parent company programmer at the ACM meeting. The latter conveyed a strong impression that our person was a second class citizen. He even referred to her working in the **Leper Colony.**"

Evaluation

ELIC was working with the symptoms of the problem, not the cause. The problem was not a wrong perception that maintenance work was less challenging. The perception is true, for the way maintenance is performed in most companies. However, as proved by the companies included in our research project, maintenance work can be redesigned to make it more challenging.

ELIC did not redesign the work; it just relocated it. The new people were productive because the work was new and interesting. That situation would be true of almost any new assignment. The length of time required for the challenge to wear off depends on either of two things: 1) the scope (degree of richness) of the job, or 2) the growth need of the individual performing the job.

ELIC apparently hired high-GNS personnel for the new subsidiary. Considering that supervisors transferred from the parent company were paramount in the selection process, such a result is not a surprise. Those persons hired personnel much like the ones they'd hire back at the parent company. The subsidiary had a higher "weed-out" rate after the probationary period because the new hires had not already been screened through completion of a degree program in computer science or information systems.

Had they known about the GNS/MPS match-up process described in Chapter 4, they could have developed selection criteria for lower GNS to match with their low MPS maintenance work. Case C discusses that selection process in more detail.

The other alternative would have been the restructuring of maintenance assignments to enrich these jobs. Of course, it wasn't necessary to form a new subsidiary if this approach was utilized. The jobs could have been enriched to match high-GNS personnel, utilizing the procedure explained in Case C.

This situation shows the expense some companies will incur to attempt to solve the problem of low maintenance productivity—without avail. It reveals the lack of knowledge in our industry on motivation theory and practice.

APPLICATION CASE G
FARMING OUT MAINTENANCE

Some organizations have contracted maintenance to firms that specialize in this work. We're not advocating that approach unless a company has excess workload. To "farm out" maintenance just because personnel prefer not to work on those jobs is costly. It is much less expensive to restructure maintenance tasks to improve the MPS of these tasks and to continue to perform the work "in-house."

It is useful, however, to examine the companies that specialize in maintenance work to determine how they achieve productivity levels sufficient to make the business profitable. The first company is Maintenance Specialists, Inc. An analysis of their results in maintenance productivity is provided below.

"Hazard" Pay

The following is the response of the president of one of the maintenance companies responding to our question about why his personnel didn't consider maintenance work routine and unchallenging. "Sure, the work is less interesting," he said. "As a consequence, I pay higher wages for doing it than these people would earn doing the same job for our clients. It's like the hazard pay earned in some professions."

He went on to say, "I don't believe these ivory-tower philosophers who say people aren't motivated by money. The people in my firm prove I'm right. They don't like maintenance work any more than my clients' programmers. They like money, though. It's their primary motivation."

He went on to say, "This is especially true of young workers. They like money even more. I have a crew of highly motivated, youthful employees who are productive."

Although he couldn't remember specific names, Russell Money, president of Maintenance Specialists Inc., was alluding to the research of people like Abraham Maslow and Frederick Herzberg, when he referred to ivory-tower philosophers. You will recall that Herzberg, the most widely cited researcher/practitioner on the subject of motivation, developed the two-factor theory. He separated motivating factors from what he called maintenance factors. The table on the following page identifies those two categories. Note the headings which further identify the two categories as dissatisfiers versus satisfiers. Herzberg's research[1] revealed that most employees are dissatisfied when their maintenance needs are not adequately met. Meeting these needs, Herzberg found, does not alone produce job satisfaction. The employees are no longer preoccupied with the maintenance factors and can concentrate on the true motivators (satisfiers).

Note that Herzberg lists pay as a maintenance factor. His research indicated that pay was not a primary motivator for most employees. Pay must be adequate or the employee is distracted from the true motivators. However, additional amounts of pay will not produce

Dissatisfiers— Maintenance Factors	Satisfiers— Motivation Factors
Company policy and administration	Achievement
Work conditions	Recognition
Technical supervision	Advancement and growth
Interpersonal relations with peers	Responsibility
Job security	The work itself
Pay	

equivalent increases in productivity, according to the Herzberg research.

Herzberg's research was replicated for data processing by Jac Fitz-enz.[2] In 1978, Fitz-enz surveyed 1,500 programmer/analysts, project leaders and managers. In surveys conducted by Herzberg, pay ranked 6th place in importance to workers. In contrast, DP professionals ranked it 10th place in importance.

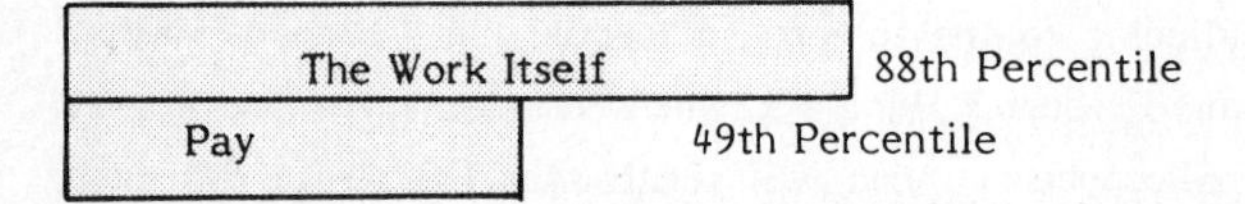

Relative Importance of Pay Compared to the Work Itself
(Data From Fitz-enz Study)

Then, how does Russell Money get his employees to produce well by paying more than the going wage? Is he misleading us?

No—he is probably right. Unintentionally, he is replicating another company's experience where similar results were produced. It was Lincoln Electric Company (the real name of the company) of Cleveland. We will analyze below the two firms and attempt to explain the apparent refutation of the Herzberg motivation theory.

Productivity for Pay—Lincoln Electric Experience

During World War II companies with unusually good productivity records received a great deal of attention. It was hoped that their methods might be introduced in other companies to produce a direly needed increase in U.S. plant output.

One such company was Lincoln Electric. Productivity per hour was very high. James Lincoln, company president, was a colorful fellow and was widely quoted in publications including Time and FORTUNE. He attributed the high productivity to a pay incentive program that he had introduced. His workers were earning an average of 50% more than their peers in other Cleveland companies. They were almost 300% more productive than their competitors. Absenteeism was one-fourth that of competitors and turnover was extremely low. Lincoln was convinced that people are highly motivated by financial incentives —and his company's records substantiated his beliefs.

Several other companies introduced similar programs; however, the results were disappointing. While slight improvements occurred, none were anywhere nearly as significant as the Lincoln results.

Analysts were assigned the task of determining the reasons for this discrepancy. The result of their analysis is an integral part of the motivation literature and the Lincoln Electric Company experience is considered as a classic case in teaching motivation concepts.

In addition to careful evaluation of the mass of data accumulated, the analysts interviewed personnel in the companies. They confirmed James Lincoln's convictions that his workers were highly motivated by financial incentives. But they also determined that Lincoln's employees were not representative of the general work force. Their results are best summed by the comment of a union leader who said his union had never seriously tried to organize Lincoln Electric. Concerning persons willing to work diligently for financial incentives, he said, "Of course, there are those who have tried it, and found out that it just wasn't the kind of life they want to lead—and that kind quits. But then, there are those who like that kind of thing and all the things you can buy with money, and they stay. So what happens is that Lincoln Electric attracts the kind of people they want—people who can and will work at a terrific pace for money. The other kind of people—the ones who want to enjoy life and to enjoy their work and maybe not make so much money—that kind doesn't last at Lincoln. You might say it's sort of a process of natural selection. Lincoln is all right, though. But you couldn't pattern all of American industry after them. A lot of businessmen around Cleveland say J. F. Lincoln is eccentric. He isn't. He's fair and honest, and he's a bear on efficiency. But I think he puts too much store by monetary incentives—but then, there's no denying he has attracted people who respond to that type of incentive."[3]

In an area as large as Cleveland, it is possible to attract that small percent of the worker population which places high priority on pay. The percent is small, however. Indirectly, the Lincoln Electric case confirms the Herzberg theory of motivation.

Now let us look at Maintenance Specialists Inc. with the insights gained from the Lincoln Electric Company case.

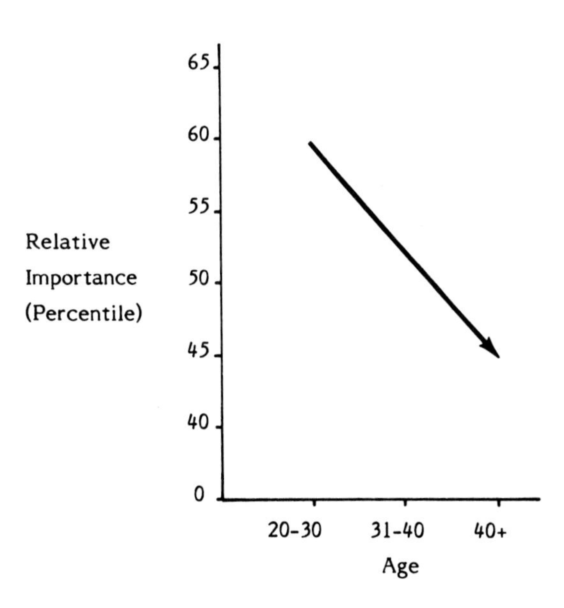

Declining Importance of Pay Related to Age
(Data from Fitz-enz Study)

Maintenance Specialists Inc.—Productivity for Pay

Russell Money's company is located in New York City. It has less than 150 employees. That number represents an infinitesimal percent of the New York City DP work force. Even when the other New York maintenance firms are added to this number, it is inconsequential. These firms are able to attract that small percent of the DP population for which pay is not in 10th place in importance. The Fitz-enz research identifies that the work itself is clearly a primary motivator, in the 88th percentile for most programmer/analysts.

Russell Money's opinion on young workers being especially influenced by financial incentives is not invalid. Fitz-enz data shows a correlation between pay-rank and age. The correlation is inverse as would be expected. Nevertheless, the difference is inconsequential. As age increased, the importance of pay decreased in almost a straight line relationship, from the 59th to the 52nd to the 46th percentile, and from ninth to eleventh in rank order.

Maintenance Companies Successful for Other Reasons

The other maintenance companies typically utilize another approach. Those companies where management is knowledgeable concerning motivation theory utilize the work redesign procedure described in Case A. They enrich the job to match the GNS of their employees, which are not unlike the employees in the clients' DP departments.

They can't change the work itself; they must handle it the way the contract specifies. They can change the way the work is performed, however. For example, they can emphasize goal-setting activities so that employees can have a great deal of autonomy. Performance standards are emphasized with sound reporting procedures to insure feedback. Personnel are assigned accounts (clients) and work with them, from specification of the work through implementation of the modified system—insuring high task identity. Skill variety may be constrained by the type of work being performed—with little opportunity for enhancement. Task significance may also be difficult to provide because the maintenance personnel are isolated from the user—they only communicate with client liaison personnel. Nevertheless, enhancement of three of the five core job dimensions enables MPS to be enlarged. A better GNS/MPS match is achieved, resulting in good productivity.

Conclusion

In most cases it is uneconomical for a company to farm

out maintenance. It can enrich its maintenance jobs to make the work challenging; the resulting productivity should be satisfactory, making it uneconomical to contract to another firm.

If a company has a temporary backlog of work and does not want to hire temporary personnel, farming out work is a reasonable alternative. If you have to farm out work, it makes sense to keep the new development work and release the maintenance work. Unfortunately, it is less complicated to farm out the new development work. The specifications are up-to-date and the package tends to be more portable than maintenance of an existing system. So, some companies contract out the more interesting work and keep the more-difficult-to-motivate work for existing employees. Nevertheless, the remaining maintenance work can be enhanced, as illustrated in previous cases. It may be easier to enhance the maintenance work than to package it for contracting.

There is another case where it may be economical to contract maintenance work. However, the situation is unique. Some organizations are experiencing stringent personnel restrictions due to the recession. Some organizations have put a freeze on hiring. Others have a headcount control. Others have a salary freeze. Managers of these organizations have little flexibility in circumventing these controls because they are carefully monitored by company control groups.

Paradoxically, some of these managers have budget to contract software, including individual or personal contracts. The results are somewhat bizarre. In some organizations we observed contractor personnel sitting next to company personnel, performing similar work—but with great pay disparity. The manager pays a contract fee which is always higher than internal pay scale. Word gets around and company morale plunges. Although this situation is grossly inequitable, DP management has few alternatives; unrealistic company policies have brought about this situation. DP has a heavy backlog of work and this is often the only viable approach within company policy.

In two of the companies we surveyed, creative managers lessened this negative impact by the way they assigned work. They made sure company employees were assigned the most interesting and stimulating tasks. The less challenging work was contracted.

Herzberg's research would indicate that the pay inequity still causes conflict and less-than-full productivity. That theory proved true in these companies. However, turnover lessened when employees were assigned the best work. Also, their productivity was good. This result substantiates the Herzberg and Fitz-enz findings—pay has a much lower priority than the work itself. So long as employees have adequate income, the challenge of the work will more than offset the inequity in pay for contract personnel.

References

1. Herzberg, Frederick, The Motivation to Work, John Wiley and Sons, Inc., New York, 1959.
2. Fitz-enz, Jac, "Who is the DP Professional?" Datamation, September, 1978, pp. 125-128.
3. "Observations on the Lincoln Electric Company," The Administrator, John Desmond Glover and Ralph M. Hower, 1963, Richard D. Irwin, Inc., Homewood, IL, pp. 243-288.

APPLICATION CASE H
SPREADING MAINTENANCE AMONG ALL PERSONNEL

Sally Cameron, one of the few female DP vice-presidents in the country, established the new maintenance policy at General Equipment. She had discussed the maintenance motivation problem with peers at a national meeting of the SMIS (Society for Management Information Systems). Several had indicated that they were having success by allocating a portion of each programmer/analyst's time to maintenance. One cited an SMIS presentation by Couger which explained the significantly lower MPS for employees assigned a high proportion of maintenance. The accompanying figure shows the difference between the national norms and the employees in the motivation research project. Those where maintenance activity accounted for more than 80% of their work had much lower ratings on the five core job dimensions. (See figure on following page.)

"Why saddle a few people with a preponderance of maintenance work," one vice president of DP said. "Spreading the less desirable work across the board is a much more equitable approach."

This approach seemed reasonable to Sally. Upon return, she called her managers together and said she would like to establish such a policy and asked their reactions. As might be expected, the responses ran the gamut. But after two hours of discussion, the consensus was to adopt the policy. The feelings of most of the group were exemplified by Craig Hopko's comment, "We might as well try it—what we've done so far hasn't been successful. Our productivity definitely needs improvement."

Contradiction of Motivation Theory

Was this policy a contradiction to motivation theory? Not necessarily. If a company chooses to perform maintenance without work redesign, such an approach is probably the best solution. The core job dimension of skill variety is enhanced, because people are working on two very different types of activity. They are also able to use a variety of skills. They may be required to continue the programming procedures associated with the original design methodology in maintaining an existing application. At the same time they will typically be using a different set of tools/procedures on the new applications they are developing.

Feedback may be enhanced because maintenance is more quickly completed and results are known much sooner than for new development work.

On the other hand, the other three core job dimensions might be diminished enough to offset the enhancement of skill variety and feedback. Task identity may be lessened because the work is spread between too many persons. It may be difficult for each person to relate his/her part to the whole. Task significance may be lessened because the typical view of maintenance is that it is less significant than new development work. The very act of spreading maintenance work across the organization implies that it is a necessary evil. Autonomy may be lessened because of the need for stronger coordination when the maintenance activities are spread among many persons. If a person had

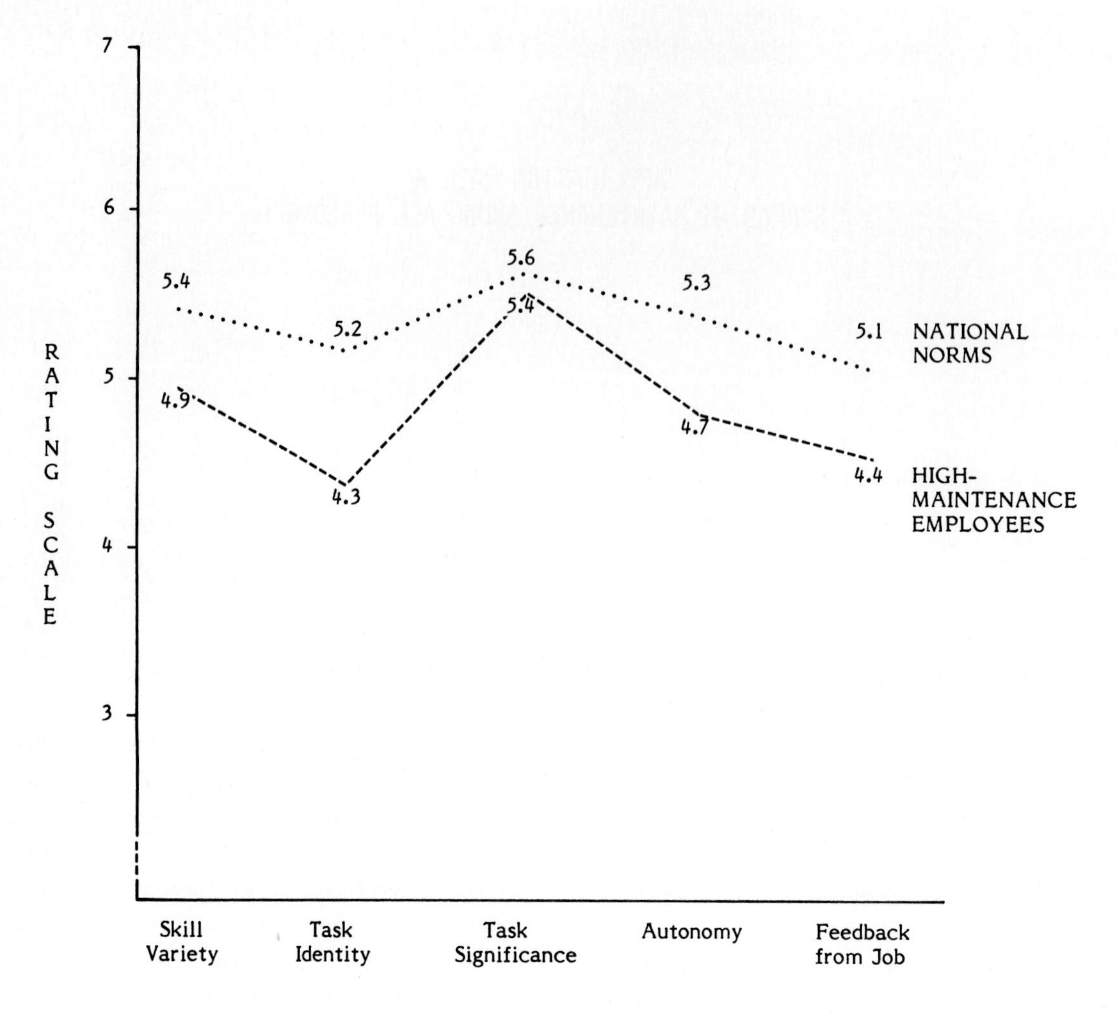

the complete responsibility for maintaining one application, he/she could be given much more autonomy. Several persons dividing that responsibility would have constant coordination, lowering the opportunity for autonomy.

The final consideration is sheer volume of maintenance. It occupies more than 50 percent of the programmer/analyst labor budget in the typical organization. Under the new policy all personnel would have 50 percent of their work devoted to maintenance; it would lower average MPS for each person. Assigning some persons more and others less than 50 percent maintenance causes an inequity that will be a continuous "thorn in the side" for management.

In a company where maintenance is twenty percent or less of the workload, such a policy as Sally Cameron's might be viable. Something must be done to change the lack of balance of CJD (core job dimensions)—as shown below, skill variety and feedback were enhanced but task identity, task significance, and autonomy were lessened.

The following illustration shows the effect of such an approach. The balance is tipped from positive to negative such that average MPS is lowered for the department. This will probably result in a mismatch of GNS and MPS.

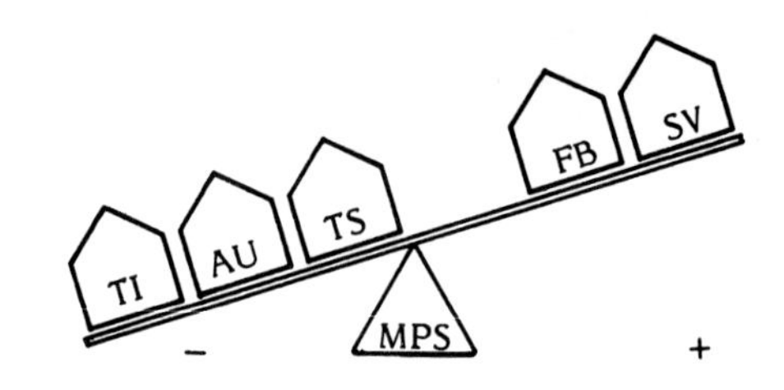

By concentrating on enhancing one of the CJDs on the left side of the scale, MPS may be as high for the maintenance portion of a person's job as it is for the new development portion.

Task significance is the easiest to improve. By insuring interaction with the users of the application be maintained, the programmer/analysts should gain a feel for the importance of the work to the company. Several activities as described in earlier cases (i.e., Cases A, D) would need to be undertaken to insure that such a result occurs. With task significance enhanced the result could be that illustrated below. Enough improvement occurred in enhancing core job dimensions to offset the lower task identity and autonomy. The overall result is an increased MPS that matches GNS and that results in satisfactory motivation and productivity.

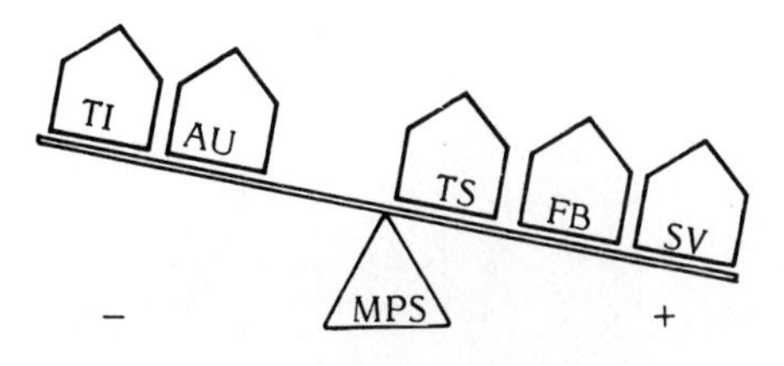

Will The Policy Work Where Maintenance > 50 Percent?

But what would need to be done for the policy to work in the typical company—where maintenance occupies much more than 20 percent of the programmer/analyst activity?

To avoid a lower MPS for all employees, a work redesign effort would be required. As illustrated in Case B a bottom-up approach where all employees are involved in suggesting activities to enhance CJDs produces the best results. By their participation, they are more committed to the work redesign process and to implementing the agreed-upon changes.

Then the question arises, "If we are going to the trouble of work redesign, why reallocate the maintenance tasks to all personnel? Shouldn't we just leave it as presently assigned and enhance the maintenance jobs where they are presently assigned?" Absolutely. That way the assignment of maintenance is performed in an optimal fashion, by one or two or however many persons necessary to naturally accomplish the scope of responsibility involved. The problem of coordinating the work when it is unnaturally split among many individuals is a significant one.

As shown in earlier cases, establishing the policy to "spread the necessary evil" is treating the symptoms and ignoring the cause. The cause is a low MPS activity that needs enhancement; the work redesign procedure is the appropriate resolution.

The following figure shows the results in one of the firms where the researchers assisted in a work redesign project. The columns on the right reflect the situation with maintenance being performed in traditional ways. The columns on the left reflect the enhanced maintenance jobs. A good GNS/MPS match resulted from the work redesign effort.

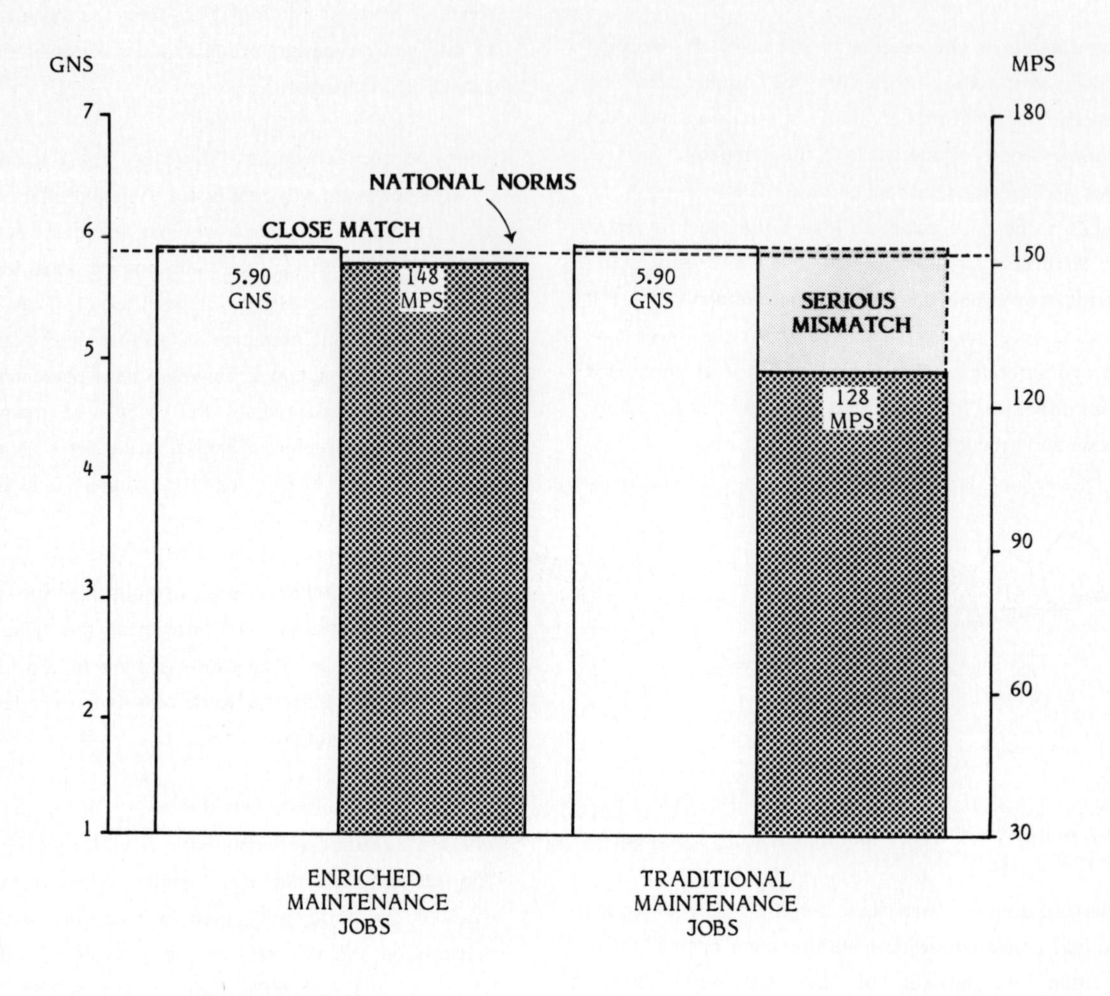
GNS
MPS
7
6
5
4
3
2
1
180
150
120
90
60
30
NATIONAL NORMS
CLOSE MATCH
5.90
GNS
148
MPS
5.90
GNS
SERIOUS
MISMATCH
128
MPS
ENRICHED
MAINTENANCE
JOBS
TRADITIONAL
MAINTENANCE
JOBS

APPLICATION CASE I
REDIRECTING TOP EMPLOYEES' INTERESTS TO MAINTENANCE ACTIVITIES

Jim Winkel, General Equipment programmer, told his boss, "If you assign me a maintenance job, I'll quit." This was not an idle threat—even in recessionary times when jobs are not nearly as plentiful as previously. Although he reacted to the ultimatum as most supervisors would, Craig Hopko submerged those feelings and responded as positively as possible, "I'll try to continue to give you as much new development work as possible, Jim, but the company policy is to spread the maintenance work among all employees. I can't assure you that you won't be assigned some in the near future. Others feel similar to you about maintenance and it would be inequitable to single you out as the only person not assigned some maintenance work."

"Look, Craig," Jim replied, "You know I'm one of your most productive employees—I have a special ability to optimize programs. My programs end up with the least amount of code to accomplish the required functions and keep CPU usage at a minimum. Why misuse this capability by assigning me to maintenance—surely company policy should not take precedence over efficiency."

"Your point is well taken, Jim. Let me think about it a while. I'll hold off the maintenance assignment for the meantime."

As Jim left his office, Craig pondered the issue. "If I let Jim get away with no maintenance assignment, I'll take a lot of heat from the rest of my people. On the other hand, I have to agree that he is my best programmer. Should I siphon off some of that talent into maintenance when I've got critical new applications schedules to meet?"

After examining this issue from all angles during the next week, Craig hit on a possible compromise. "Why not try to redirect Winkel's optimization capabilities for new applications to a similar need in maintenance. We have some programs that are CPU hogs that could use his special talents."

Redirecting the Exceptional Programmer Toward Maintenance

Craig asked Jim to meet with him to discuss the issue. "Jim, I've thought your request over carefully and have an idea which is more than just a compromise. I think it will fit in with your objective of challenging work and at the same time will facilitate efficiency of operations. We have some applciations that are not as efficient as they should be. They require a great deal of central processor resources. With your ability to optimize systems, I think you could modify these programs to be much more efficient. At the same time you could incorporate changes that our users have requested to expand the capability of these applications. How about taking this challenge?"

Jim mused over the suggestion for a few minutes. He drummed on the table with his pencil eraser as he thought.

"My first impression as you were talking was that you were trying to manipulate me. You could tell the other managers that you'd forced me to back down if I accepted your suggestion." Jim was never one to mince words. You always knew exactly where you stood with him!

"But as I thought further," he went on, "I can see that this approach really does meet with my personal objectives. It would be interesting to take someone else's program and demonstrate how I could improve it." Modesty was not one of Jim's virtues, either.

"Yes, I'll give it a try," he said. "Bring on your toughest job and I'll tackle it."

Results

The outcome was better than expected. After the first assignment proved successful, Jim went on to a second application where response time was a problem. A large on-line application, with more than 100 terminals in an order entry area, irritated operators due to waiting time for the system to respond to each entry. Jim was able to cut average response time from 8 seconds to 5 seconds for each transaction. Although the improvement might appear inconsequential because of savings of only a few seconds per transaction, operators' complaints on response time were almost totally eliminated and productivity increased significantly.

The only problem was the feeding of Jim's huge ego. He was inclined to make statements such as, "I can take anybody's program and cut running time by 25 percent." Nevertheless, the overall results were positive.

This may appear to be such an exceptional situation that it is inappropriate for recommendation to the typical DP organization. That is not the case, however. Every department has 5-10 percent of its employees in the Winkel category. Their impact on maintenance productivity far surpasses their percent of the labor budget. Improvements range up to a factor of 10. Granted, the overall departmental effect is less than that factor, because you have transferred this optimization advantage from the new development area to the maintenance area. With only 1 in 10 or 1 in 20 of your personnel having the Winkel-type capabilities, they are a scarce resource that needs to be assigned optimization tasks in both maintenance and new development.

APPLICATION CASE J
THE PRODUCTIVE TROUBLE SHOOTER

One of the persons interviewed in this research project was sullen and uncooperative. He would answer questions with a "yes" or "no" and avoid elaboration if at all possible.

After the interview his supervisor shed some light on the problem. "Chuck has been with the company over 10 years," he said. "He started as an operator and worked his way to the top. He was a trouble shooter. Although he had only a high school diploma, he clearly was intelligent and made good use of his talents. He worked second shift and taught himself enough programming that he could make quick fix-its late at night when a programmer was not available to correct the problem to get a daily job back in production in time to meet the output schedule the following morning."

The supervisor concluded, "Maybe we made a mistake in transferring him to programming early this year. However, we had a reorganization which called for anyone doing programming to be moved back under the programming manager in the development department. Both the systems programmers and trouble shooters had previously worked for the operations manager. The move was justified on three points:

1) Programming standards could more easily be enforced if all programmers worked for the same boss.
2) Programming job positions would be more equitably distributed.
3) The programming career-path would be enlarged for those personnel coming over from operations. For example, the trouble shooters were at the top of their career path. Now they are at the Programmer II level and have the whole range to progress through."

When we asked about the type of work now assigned to trouble shooters, the supervisor responded, "We still have the trouble shooter job in operations. They are not allowed to touch the code, however. They must call a programmer when it is the program that caused the problem."

"What are the former trouble shooters working on?" we asked.

"Most have been assigned maintenance jobs," the supervisor replied. "Some will be assigned to new development when they are better trained and have proven themselves."

"What do you feel about their productivity so far?" we asked.

He shook his head in discouragement. "It's not good. We're into our 7th month since the change and these fellows are not enthusiastic about their work nor very productive. They don't interact too well with the other programmers and tend to spend their breaks and lunch times with other former trouble shooters."

Apparent GNS/MPS Match-Up Without Productivity

We discussed this problem with another supervisor who was familiar with our work and commented, "Based on your findings that operations personnel have lower GNS than development personnel, it would appear the GNS of these former trouble shooters would match well with the lower job scope of maintenance work."

We agreed. "Then," he asked, "why aren't we getting satisfactory productivity?"

"We'll interview other trouble shooters and get back with you," we replied.

In discussions with other trouble shooters we concentrated on the things liked and disliked about the present and previous job. The consensus opinion was that the new jobs were not nearly as interesting. They enjoyed the ability to solve a tough problem that other operations personnel couldn't resolve. They didn't mind the pressure to reach quick resolution—in fact, they thrived on it. They didn't mind working overtime when the problem couldn't be resolved during their regular shift.

Conversely, they felt maintenance work relatively unchallenging. They also sensed that the other, college-trained programmers felt superior. "They'll hardly give me the time of day," one trouble shooter said about his new co-workers. "I work hard but want to break the routine throughout the day to occasionally discuss non-business things. They'll talk to me but it's obvious they think their work is more important and make it clear after a few minutes that they would rather be working than talking to me."

Different Personality Characteristics of Trouble Shooters

Fortunately, this problem was identified in one of the first organizations in the research project. We continued to explore it in subsequent organizations.

It became apparent after a number of interviews that trouble shooting activities appealed to a small percent of personnel in DP departments. Most programmers and analysts prefer the longer term projects and would not respond well to the constant pressure of the trouble shooting role. Most of them do not object, in fact respond positively, to high pressure, high overtime situations occasionally in new development work. But they would not like a "steady diet" of such activity.

There are exceptions, however. For example, some system programmers experience the same kind of excitement as operations trouble shooters when confronted with tough, short duration problem solving situations. So do some of the applications programmers and analysts.

We've not researched hardware maintenance personnel but are told by other researchers that the same situation exists for that job category.

It is clear that people with such a personality make-up are in the minority, however.

MPS of Trouble Shooting Tasks

Trouble shooting tasks as described above have most of the ingredients of high scope/high MPS work. Task identity is high because they work a problem from beginning to end. Task significance is high because only senior people with a great deal of job knowledge are qualified to be trouble shooters. High visibility occurs from completing these problem resolutions in short order. Feedback from the job is quite high—it is frequently non-ambiguous.

Therefore, a high GNS person matched against the high MPS of trouble shooting should be expected to be productive.

Nevertheless, other high GNS persons would not perform well under the continuous trouble shooting environment.

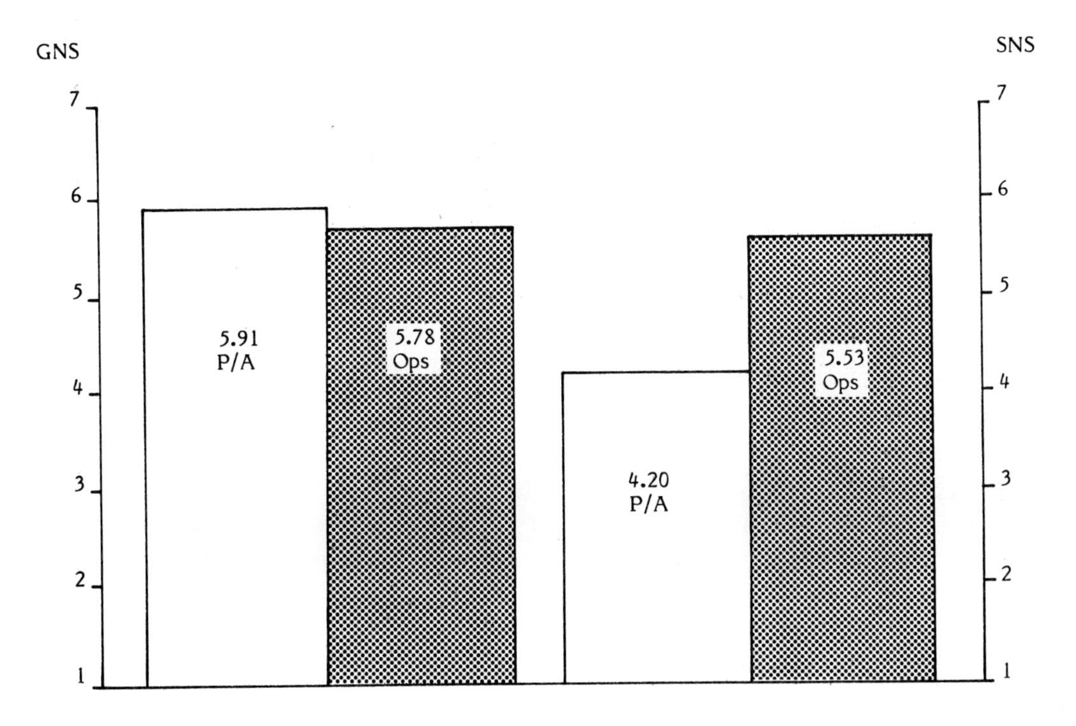

Comparison of GNS and SNS Between Programmers and Analysts, and Operations Personnel

So, this affinity for short duration work is another variable among those that must be considered by supervision in task assignments.

Reason for Low Productivity of Former Trouble Shooters

When these trouble shooters were transferred to the programming department—just because the previous resolution procedure allowed them to make changes to the code and therefore identified them as quasi-programmers—a major change in status occurred.

They were the elite of the operations department—both in rank and in prestige because of their special ability.

They moved to "bottom of the totem pole" in prestige and rank when they were transferred to programming.

The opportunity to move to a rank and pay scale higher than the colleagues they left behind in operations did not compensate for this reversal of status.

Effect of Difference in Social Need Strength

The transferred trouble shooter perception that the other programmers viewed them as inferior may have been correct. However, another factor is probably more important in causing that perception.

People in operations have much higher need for social interaction. The earlier Couger-Zawacki research identified this difference. Programmers and analysts have the lowest SNS (Social Need Strength) of any of the 500 occupations studied previously. SNS of operations personnel is close to the mean of those other occupations.

When the former trouble shooter spoke of his difficulty in keeping his coworker's interest in discussion of non-business subjects, the lower SNS is probably the principal cause. The response could be interpreted as lack of empathy or as a feeling of superiority, when in reality it was not that at all. Persons with low SNS just have low satisfaction from small-talk.

The figure on the previous page illustrates the similiarity in GNS but extreme difference in SNS between operations and development personnel.

The importance of this disparity in SNS is that it is one more factor to keep the former trouble shooters from feeling compatible with their new environment.

Result

The result of such a reassignment is three fold:

1. The employee with the highest status and rank in one organization suddenly finds himself in a role reversal—near the bottom in rank and status in the new organization.
2. The employee who has been productive on short duration work now finds himself assigned to long duration work.
3. The employee who enjoys social interaction finds himself among a group of employees whose need for social interaction is significantly lower, relative to his own..

It is of little wonder that these people were not productive in their new assignment.

APPLICATION CASE K
A PROBLEM OF COMPENSATION

This case involves a large state data processing department, a central DP group. As such, it functions in an environment where a set of strict organizational rules and regulations must be applied before individual management style can be exercised. The department operates with a group of about fifty data processing professionals who are classified as programmers, programmer/analysts, and analysts. About half of the professionals are classed as programmer/analysts with about 15 programmers and 10 analysts. Multiple levels of classification exist at each of the three major levels.

While the management style of the department is strongly supportive, the overriding factor in the organization is the existence of a strongly controlled personnel system. This system affects the shop in two primary ways. First, it limits the pay scale to levels significantly under those of surrounding data processing shops. Secondly, it fixes the entry and promotion procedures for all personnel. New personnel must be hired into a six month trial period. During that time, management can release individuals without major effort. After the six month trial, employees begin an almost automatic set of periodic promotions at a frequent rate. The employee now has job security and dismissal becomes very difficult. However, the pay scale never becomes competitive with the private sector.

The data processing function is organized along functional system boundaries. Each major system has one or more senior analysts assigned to it, along with a support staff of lower level personnel. The general philosophy of the shop is that each functional group must engage in ongoing development and maintenance of its own systems.

Because of strict salary structures, the DP Department cannot compete for professionals on a salary basis. In order to meet staffing requirements, management has implemented the following hiring/training/promotion procedures.

New employees are usually not data processing personnel. Instead, the shop seeks individuals who have four year degrees in areas of study which do not offer strong job markets. In other words, they find people who are educated but underemployed and offer them an opportunity to enter a new professional field without having to incur the cost of additional formal education. The organization then accepts full responsibility for the training of the new employees.

Because of their past underemployed status, new employees find themselves receiving significantly higher salaries than they have been able to attract in the past. Even the relatively low pay scales are sufficient to attract these people. Then, as the training process (which begins immediately) begins, the employees find themselves learning new skills and applying them very early in their tenure at the shop. Newly learned programming skills are applied when senior personnel turn over coding tasks to the trainee. As more skills are acquired, the employee accepts greater responsibility for expanded tasks.

As a result of this process, new employees attain early salary satisfaction and early skills acquisition and use. Each incremental increase in skills is met with matching incremental increases in the scope of job assignments. In addition, the almost automatic promotions and raises contribute to the ongoing feeling of growth and recognition. At the same time senior people are relieved of many routine tasks by the junior employees, freeing them to tackle ever more complex and satisfying tasks.

This situation remains stable except for two major periods of potential problems. First, during the initial six months, the employee may be found to be unsuited for the job and terminated. Second, and much more detrimental to the shop, massive turnover is experienced after about one and one-half to two years employment. At this point, the employees have learned sufficient skills to be attractive to other shops in the area. In addition, the rapid raises and promotions begin to slow down and the number of new skills being learned on the job decreases. A large percentage of employees leave at this time, seeking higher salary, faster professional growth, and greater responsibilities. The employees who do not leave at this time tend to remain with the shop for long periods with low turnover.

The Maintenance Process
In The DP Department

Both junior and senior personnel are involved in maintenance tasks. However, while the senior people have a varied menu of development and maintenance tasks, the junior people are, in many cases, doing only maintenance. In addition, the senior people are expected to perform the analysis, design, testing, and implementation of program changes, resulting in high skill variety. The junior people often perform only the programming of the changes, resulting in low skill variety and therefore low motivating potential.

However, the new employee who has just learned programming skills does not consider programming tasks to be of low motivating potential. Such tasks utilize all of the data processing knowledge possessed by the individual and they are therefore highly motivating. In addition, the differentiation between development and maintenance is not terribly critical to these people. As one employee with about 10 months experience said, "As a programmer, it is impossible for me to tell whether a coding task is a new program or an enhancement to an existing system." While this implies low task identity, and therefore low motivating potential, the new employees apparently find other job factors more important in this environment.

As employees gain more skills, they are expected to use them in tasks almost immediately. This results in a smooth progression of skills acquisition and increased task significance and skill variety. Finally, after the major turnover cycle at about eighteen to twenty-four months, the growing professional is expected to utilize new employees in order to perform routine coding tasks, thereby freeing him or her to move on to more complex tasks.

Analysis Of Employees In The DP Department

The C-Z survey instrument revealed several interesting facts about employees in this organization. First, the growth need strength was at or above the national norms for all classes of employees. The new employees exhibited particularly high GNS.

In general, we expect to find high-GNS associated with low satisfaction in high maintenance environments. However, this was not the case here. The new employees rated their jobs as having high skill variety, task significance, feedback, and autonomy. Their evaluation of general satisfaction was at the national norm, and their motivating potential score was significantly above the national norm. The coordination of training and task assignments clearly contributed to a matchup between high-GNS and high-MPS.

Another indication of a healthy organization of the maintenance function was found in the analysis of the core job dimensions across groups with varying levels of maintenance responsibilities. In general, the MPS was consistently high for all levels of maintenance, again indicating a good coordination between individual GNS and the level of the tasks assigned. Even senior personnel seemed to find maintenance tasks motivating, due possibly to their use of junior people to aid them in the routine activities.

Summary Comments

During the first two years of employment, these people found their work interesting and challenging because they were constantly learning and applying knowledge. Skill variety was continuous. The opportunity to immediately apply newly learned skills and ascertain the results provided feedback from the job.

Contrast this situation to that of a person who is hired from a four year computer science or business data processing program. That person has been programming steadily since his/her sophomore year in college. Many have part-time programming jobs on the campus in the computing center or assisting faculty in research projects. Many others obtain part-time jobs in local industry. The challenge of maintenance activities, due to its newness, would wear off much more quickly for these persons.

By hiring untrained personnel, this DP department not only attracted college graduates at a lower salary, but also attracted people to whom the maintenance tasks would be new and challenging. Of course this raises the question of productivity of these less experienced personnel. Undoubtedly, they know fewer programming "tricks and shortcuts" and are at the high end of the learning curve shown in Chapter 1 (page 6). On the other hand, the higher motivation can compensate for lack of experience. A more experienced staff may produce less due to lack of motivation and job satisfaction. The state DP department has utilized a nice tradeoff to achieve productivity despite its severe constraints.

In summary, this organization has been able to work within the constraints of a strict organizational environment and develop a set of hiring, training, and promotion procedures which allow effective and healthy maintenance assignments. The coordinated training and task assignments seem to be a key to their successful operation.

It is interesting to note that the personnel system which has forced the development of the procedures discussed here is also responsible for the avoidance of motivation problems after the two year turnover cycle. Management of the department agreed that most of the employees would become dissatisfied with their jobs if they stayed for long periods. While the current turnover pattern is primarily due to salary issues, one manager said that, "If they stayed longer, they would probably end up leaving anyway because of our high maintenance load."

APPLICATION CASE L
THE ULTIMATE SPECIALISTS

Software Systems, Inc., is a national organization with a reputation for successful design, construction, implementation, and maintenance of large systems. In one of their many locations, a major project was completed about two years ago with the system moving smoothly into a maintenance mode. As a result of the system's implementation, the local staff had undergone a major reduction. Many of the senior professionals had moved to other development companies in the area in order to participate in continuing development of large systems. Some of the senior systems staff had stayed with Software Systems, Inc. for various reasons. As a result of the reductions in the group, a group of about twenty new employees was hired at junior levels to complement the fourteen senior professionals in the maintenance effort.

About one year after the implementation of the system and the resulting restructuring of the systems effort, the manager of the group announced his departure. As a result, Jim Anderson, a successful first line supervisor, was asked to move into the group manager position. As his manager said, "We know that this is quite a move up for you, but your interpersonal skills are strong and those are the skills we need right now. The group is well organized and functioning smoothly, so it should require only average administrative supervision." Then he added, "Our very real problems now lie in productivity. We have over thirty-five people responsible for the maintenance of over one million lines of code. While the reliability of the original system is very good, we are falling behind in delivery of enhancements to that system. We feel that the existing structure of the group is good, but we look forward to any suggestions which you might make to improve overall productivity."

Analyzing Causes Of Low Productivity

After a brief familiarization period in the new position, Jim began to concentrate on the productivity issue. Having worked closely with the programmers and analysts, he felt that he had some insights into the situation and he was anxious to tackle the problem. He had been reading the maintenance literature and was familiar with some of the job related situations which tend to result in motivation problems and he felt that the low production rate might be tied to one or more of those issues.

First, Jim reflected on the nature of the system maintenance being performed. Because of the high reliability of the installed code, fixit efforts were infrequent and instances of night and weekend work on emergency fixes were rare. In fact, most of the work was scheduled well in advance in order to support a rigid release cycle. These factors seemed to argue against the existence of staff burnout as a cause of low productivity. While impending release dates tended to force high work rates, they were predictable and well-separated. Everything pointed to a relatively well balanced workload.

Next, Jim embarked on an examination of his staff's perceptions of their jobs. Knowing that maintenance work is generally viewed as disinteresting and unchallenging, he was concerned that his employees might simply be exhibiting a low productivity as a result of their dissatisfaction with their jobs. However, in a series of informal discussions with individuals and small groups of professionals, he found that only a few of his employees were overtly negative about maintenance tasks. Because of a very real possibility that the organization might soon embark on a new large development effort, Jim was able to ask each employee, "If this new development effort comes through, do you want to be considered for membership on that team?" Of the approximately thirty-five employees, only about half exhibited any interest in the new project. Only four members of the group made strong responses, all in line with one very firm statement that, "The only reason I stayed on with the company was to get in on the ground floor on the next big project. If the new project is approved and I don't get reassigned, then I'll quit immediately."

In contrast, most of the staff members who expressed interest in the new development effort were only moderately interested. They wanted more information as to the duration of the project, the size of the development team, the reasonableness of the schedule, and other data which would allow them to make a clear decision as to the relative merits of the two jobs. Their responses were not those of people who simply wanted out of their present assignments.

The really interesting responses, however, came from the individuals who clearly indicated that they actively preferred the maintenance environment. One young woman said, "I have made a careful decision to strive to achieve a balance between my professional and personal life. I feel that I can best control my professional life by remaining in my current environment where my job's requirements are predictable. I feel that development tasks would reduce my ability to engage in other activities which are important to me." This employee represents a growing number of employees who present problems requiring new motivational approaches as discussed in Case A.

An older employee said, simply, "I don't want that rat race anymore," while another junior team member said, "I don't understand all of the interest in new development. I'm working on a very current and important system and I don't see what development tasks offer that is different than my present job."

All of these results left Jim puzzled as to the underlying causes of the productivity problems which were observable across the team. In a meeting with his manager, he questioned, "If its not burnout and if its not a simple dislike of maintenance, then what can the problem be?" After a series of discussions, Jim and his manager decided to seek detailed information from all of the employees as to their perceptions of their jobs. They chose the Couger-Zawacki survey instrument and administered it only a few weeks later. The results suprised both Jim and the management above him.

Analyzing Job Perceptions

First of all, the results were strikingly similar across the staff. As shown in the figure on the following page, skill variety, task identity, task significance, autonomy, and feedback from the job were all lower than the national norms.

Only the senior people who had participated in the original development deviated from the pattern, and that was in the predictable areas of task identity and task significance. Otherwise, the low scores on the core job dimensions were dismally uniform. The summary measure, that of motivating potential score, was among the lowest ever observed in a data processing shop. These low scores on the core job dimensions, coupled with a significantly high growth need strength, clearly showed an imbalance between growth need and motivating potential which could account for the low productivity. Jim immediately began a series of discussions with his staff in order to understand the implications of the numbers.

Working along the lines discussed in Case B, Jim organized a series of meetings in which employees were encouraged to offer suggestions which might improve their jobs in the core areas. Almost immediately, an undeniable pattern was apparent from the string of suggestions, some of which follow.

"I would like some type of report as to how smoothly my code gets through test."

JOB VARIABLE	C-Z NATIONAL NORMS	SOFTWARE SYSTEMS, INC.
Skill Variety	5.4	4.7
Task Identity	5.2	4.2
Task Significance	5.6	4.9
Autonomy	5.3	5.2
Feedback From Job	5.1	4.7
Growth Need Strength	5.9	6.4
MPS	153.6	118.8

"Could we set up some type of system which would let me know how well the users like the changes I participate in?"

"There are times when I feel that I could build better changes if I could deviate from the specs which are delivered to me. Is there any way that I can gain some freedom in the construction of the enhancements?"

Jim soon realized that the well organized structure of the group which he had inherited was the source of the problems which he was experiencing. The inherited structure was one which split the group up into teams which were individually responsible for single sets of processes in the development cycle. As a result, each person saw only a small portion of any given system change. Those responsible for analyzing changes seldom learned the correctness of their analysis. Those who designed the changes passed them to the construction group and moved on to the next effort. Construction personnel seldom received test reports, and almost no one received any information as to user satisfaction.

The problem clearly centered on the structure of the maintenance group. Skill variety was certainly not high, as each individual was charged with only a subset of the activities for which they had been trained. In addition, task identity and feedback from the job suffered greatly from the isolation which resulted from the organizational structure. Likewise, task identity was reduced due to a lack of understanding of the "big picture", and there was little autonomy due to the intertwined nature of the separate efforts. These low factors, coupled with high growth need strength, were certain to produce a poor match of employees to work. The result was poor motivation and, hence, low productivity.

Moving Towards An Improved Solution

Jim discussed these results with his manager. In response to his request that he be allowed to consider complete reorganization plans, he was told that customer satisfaction with the current structure was high and that major changes might weaken that relationship. This limitation of options sent Jim back to his office with some discouragement.

However, he soon realized that his strongest support in addressing the motivation issues lay in his own staff. After additional study of materials on job design and group processes, he began a series of meetings with his personnel in order to seek their suggestions for job improvement. As a result, he received a number of valid and workable recommendations. Many of these continued along the lines of the earlier feedback which he had received, suggesting procedures by which each employee could feel a part of the entire change process.

After considering all of the input, Jim selected a set of his senior people and called them to his office. "I would like to ask each of you to function as project leaders," he said. "As change requests come in, we will meet as a group to prioritize and assign projects. After you accept responsibility for tasks, I expect you to schedule them through the analysis, design, construction, test, and implementation groups. While you will have to utilize available personnel in each group, I expect you to have weekly meetings to discuss active projects with all involved employees. These should be short meetings, directed only at providing feedback on progress and encouraging the airing of problems. Problem resolution should occur outside of the meetings."

Discussion Of The Solution

The solution implemented by Jim involved a localized change in organizational structure. The resulting project orientation allowed employees to become more involved with the entire change process rather than being isolated in a localized portion of the effort. For the senior employees, the increase in the scope of their assignments resulted in increased skill variety and task identity. The increased richness of their jobs produced a healthy match to their high growth need strengths.

For the junior employees, the results were also positive. The weekly meetings increased task identity and supported an increased understanding of the true significance of the different system modifications. Even more important, the meetings provided feedback within the group. These improvements in the core job dimensions resulted in increased motivation and, therefore, higher productivity. As a result, subsequent releases on the system incorporated more changes then before, reducing the backlog of requests.

In all of the organizations which were studied as a part of this project, this one stood out as having the highest level of specialization among its maintenance personnel. In addition, it had the lowest set of measurements on the job variables. The lack of job feedback and task identity which result from absolute compartmentalization has massive effects on all aspects of the job. As a result of some fairly simple changes, Software Systems was able to improve its motivational potential dramatically, with corresponding improvements in productivity.

APPLICATION CASE M
A CONTINUATION OF DEVELOPMENT

One of the most consistent results of the research efforts from which this report is constructed involved the vast differences in the practical definition of maintenance in the study organizations. While most management and employee responses supported a clear understanding of the 'fixit' component of the maintenance effort, the 'enhancement' component was often a catchall category for all kinds of systems efforts which did not fit nicely into other classifications. For example, one organization defined maintenance as, "... any work on an existing system." While this is an intuitively satisfying definition, it resulted in a massive rewrite of an accounting system being classed as maintenance. In reality, only a few of the features and reports of the original system were retained, and the new system involved a conversion to an on-line, data base system from a batch file oriented system.

Another organization was experimenting with a definition of maintenance which was dependent on the total expenditure level for each project. Their rule was that efforts which cost more than $75,000 were new development and that anything else was maintenance. In yet another organization, the management was firm in its separation of maintenance and development. "It's simple," said one supervisor. "If it's on the pink work order sheet, it's new development, and if it's on the yellow sheet, its maintenance."

These examples (and many others) illustrate the vagueness of the definitions of maintenance and developments in the industry. The resulting effect on the motivational environment is complex.

In general, we expect development tasks to exhibit more richness than maintenance tasks. This implies that the performance of development activities under the maintenance classification should increase the motivation potential of the affected jobs. However, a number of conflicting factors affect this situation. Some of these factors include the maintainability of the system, its age, and its structure.

Development vs. Maintenance

In general, it appears that most organizations tend to classify development activities under maintenance far more often than the converse. While there are many reasons for this apparent situation, the most commonly expressed explanations involve the organization's perceptions of the two types of activities. For example, one manager stated, "Whenever we get a formal development project, we get a budget and time-table right along with it. If we violate either the budget or time, then people notice it very quickly. However, the maintenance costs are treated almost as a special type of overhead for the department. Extra costs which would be highly visible in the development budget are simply absorbed into the maintenance budget without a lot of comment."

Another manager put it even more bluntly. "The literature tells us that nobody knows how much we're spending on maintenance activities and we pretty much accept that. So it's just easier to dump small efforts into that account and let them get lost than it is to go through a more formal process for development efforts."

For whatever reasons, many maintenance personnel are, in reality, doing developmental work at various levels. The mix of maintenance and development may, depending on the specifics of the situation, create either a good or a bad motivational environment. In general, these situations require consideration of both the motivational issues and the specific maintenance problems associated with the task. The following three situations illustrate the vast differences associated with the organization and mix of maintenance and development.

Development, By Any Other Name . . .

ABC Corporation, a manufacturing organization with several hundred persons in the systems department, appeared at first to be similar to many of the other organizations discussed in this report. Their development backlog was fairly constant at about two years, and systems development was progressing at a steady pace with new systems being implemented regularly. Maintenance was performed by a separate group.

The director of the department, however, expressed serious concern about the attitude of the maintenance personnel. "We have one of our best managers runnning that group," he said, "and some of our best supervisors and personnel, too. Yet they are almost to the point of rebellion. If there is an identifiable motivation issue which we can address, then we need to do so immediately."

In discussions with the maintenance manager, the root problem was identified immediately. "We continually get new systems delivered into maintenance with massive problems," he complained. "Sometimes the development group even passes undelivered systems to us. They run out of time and money and declare by default that development is completed, leaving us to finish testing and debugging, and handle the implementation."

The professionals in the maintenance group voiced similar complaints. Some of the comments were:

"We have to handle the worst of the development group's problems. The integration errors are the worst to find and fix, and we have to solve their problems for them."

"We have no part in the design of the systems which come to us, but we have to make them work."

"We're scapegoats! We look responsible for poor implementations of other people's designs, and our budget suffers while the development teams move on to new projects."

Further study of the ABC Corporation maintenance group showed that maintenance tasks were well organized and that the maintenance job, itself, was well designed from a motivational perspective. The severe problems with the group were tied directly to the perception that new system problems were being passed to them inequitably.

This case is not one which can be addressed directly through the examination of core job variables and subsequent job redesign within the maintenance group. The problem here involves the organization and operation of the systems group itself. An apparent effort to streamline the development operation had resulted in a massive negative motivational impact on the maintenance group. Only a careful examination of the systems group as a whole can effect real change here.

Some organizations have created healthy environments in this type of situation. The most progressive groups consider the problem of maintenance at the very beginning of the project. A maintenance person sits on the development team just as a user representative is included. The maintenance professional continually evaluates the design for testability, generality, and understandability. In certain cases, organizations have even required that the maintenance group formally sign off on new systems before acceptance. Just as users have the right to accept or reject systems from an applications perspective, these organizations require formal maintenance signoff.

There are two major strengths to this approach. First, the organization, by formally recognizing the importance of maintainability, will save money over the life of the system. This is an important objective since the cost of maintaining the system will be at least twice its devlopment cost. Second, and more important to this report, this approach gives maintenance personnel an active responsibility for certain features of systems which are to be delivered to them for maintenance. This creates a clear and unambiguous set of responsibilities for both development and maintenance personnel and clarifies the boundary between development and maintenance. Motivationally, this role clarification provides for less ambiguity and confusion and thus more autonomy.

The Problem Of Documentation

Another problem which is closely tied to systems development prior to delivery into maintenance involves documentation. Despite the current emphasis on the development of documentation as a byproduct of systems development, many systems are still being delivered without supporting documentation. This situation exists for systems being delivered today, and it is much worse for older systems. The effect on the maintenance function is clearly negative from two aspects. First, poorly documented or undocumented systems are difficult to modify. Second, a lack of documentation can create an ambiguous job environment with respect to maintenance. Many shops fail to clarify the responsibility of the maintenance programmer when he/she is faced with a lack of documentation.

Consider the situation where changes must be made to a poorly documented system. In order to make the change, the maintenance programmer must first bound the change area and study the affected code sufficiently to understand its function and predict the effects of change. The individual usually creates notes and other informal documentation during this process, thereby creating a form of documentation for the affected code.

This process, while necessary to the maintenance task, is often overlooked in the estimation and management of maintenance. In reality, it is a set of tasks which are left over from an incomplete development process. Yet, because it so often occurs during maintenance, this informal documentation loses its identity and therefore creates ambiguity with respect to the time needed to perform a given change.

The figures on the following page clearly show a difference in perceptions between management and employees in the relationship between task size in lines of code and task time. While management and professionals agree that most fixit tasks are small, they disagree on the distribution of task time. Management feels that task time is related to task size, while employees see task time as relatively independent of task size. In discussions with professionals, this difference is often ascribed to the fact that changes of one line of code often require as much analysis and documentation as much larger changes.

When documentation must be performed as a part of maintenance, it must be recognized as a subtask so that time can be allocated to it. In addition, if the documentation is created, the organization may want to consider creating a standard by which that material is formalized and integrated into a documentation package. In this way, existing documentation can be kept current and systems with no documentation may be documented, a piece at a time.

From a motivational perspective, a formal recognition of the documentation task reduces ambiguity through a clear definition of the maintenance effort. While documentation may not be every professional's favorite activity, the organization's statement of the importance of the activity, supported by a policy which assures that the task is completed, may increase task significance and improve the motivational characteristics of the effort. More important, the organization which makes statements about the importance of documentation, but which fails to enforce the standard or provide time and budget for its implemetation, will automatically reduce perceived significance of that effort.

Modify Or Enhance?

The final maintenance/development issue considered in this case involves the enhancement of sections of existing systems as a natural extension of making modifications. This issue arises naturally on old systems which are unstructured and which contain logic which is not sufficiently general to support current modifications

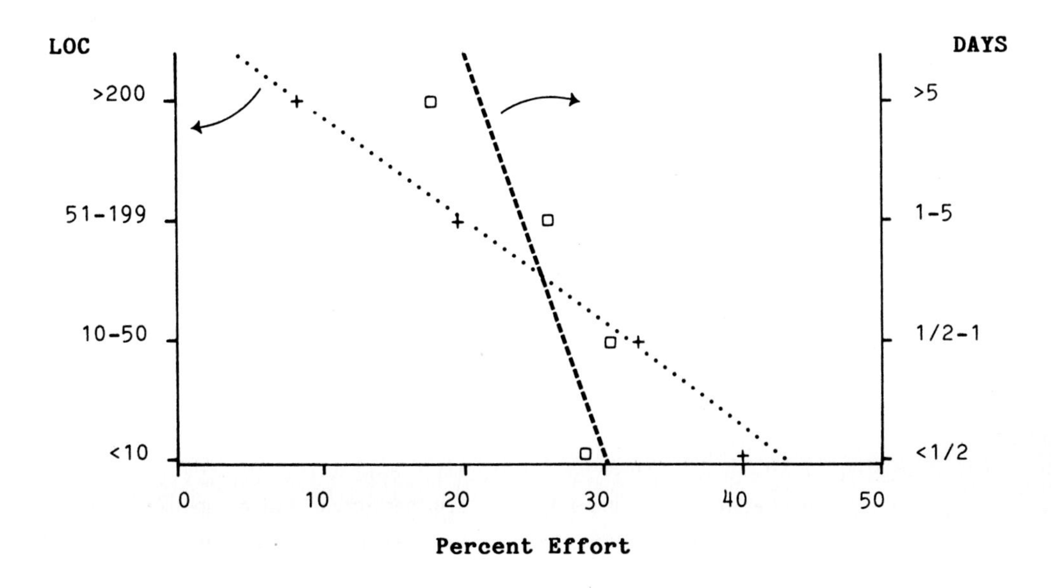

EMPLOYEE PERCEPTIONS: SIZE VS. TIME

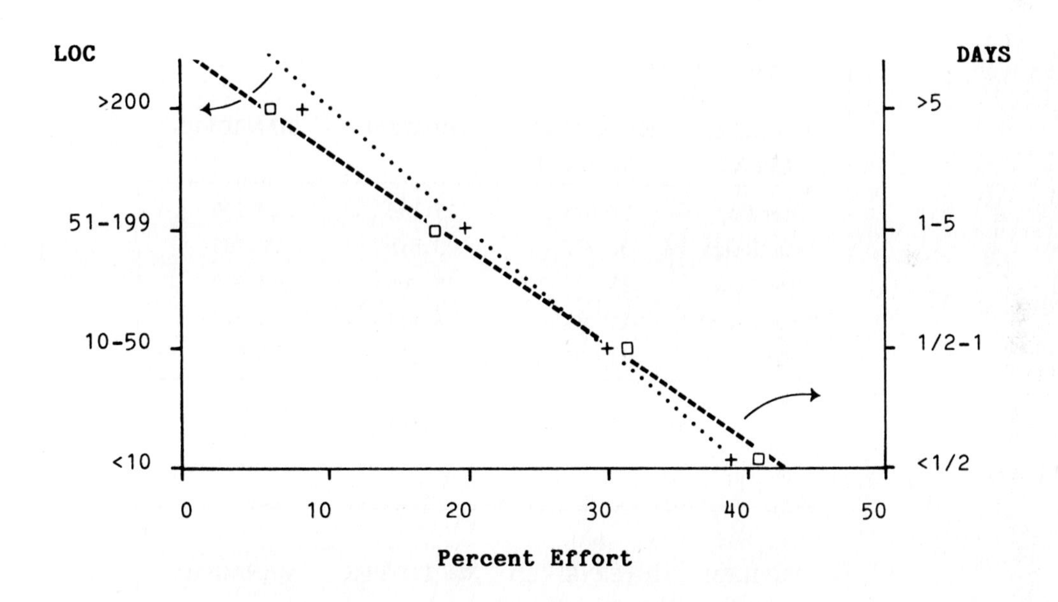

MANAGEMENT PERCEPTIONS: SIZE VS. TIME

easily. In many situations, the argument for rewriting a section of an existing system is presented most effectively by the common cry, "It would be far easier to rewrite this than it will be to change it!"

While the rewriting of sections of code during maintenance contains some elements of development, it differs from the problems of incomplete development and documentation in a very important way. The need to enhance sections of code is best determined as a part of the maintenance effort and the process of enhancement is often naturally included in that same effort. Therefore, many organizations view system enhancement and modernization as an integral part of the maintenance group's responsibilities. The differences in motivational effects lie primarily in the manner in which the organization handles this aspect of software maintenance.

As noted in Case A, some systems are not only old, but also in the process of being replaced through an existing development effort. In that situation, efforts to enhance the old system are clearly not cost effective unless a strong argument can be made that a local rewrite really is faster than a change. The situation of more interest here, however, is that of the existing system which is expected to remain in place for a reasonable period of time. This may either be an old system with an extended useful life, or a relatively new system which has been subject to rapid changes in its operations environment or applications requirements.

These systems must be maintained effectively in order to ensure an extended life, and the decision whether to modify or enhance is one which not only affects the software, but the maintenance staff as well.

In one organization involved in this research, a firm rule applies to all maintenance, forcing all changes to be made within the logical structure and code of the existing system. The maintenance personnel involved in that shop make frequent comments about being constrained by the old code. These comments become particularly relevant during discussions of autonomy, with one representative comment being, "There's not much room for any kind of originality when you are constrained to fit into existing bad code." Another experienced maintenance programmer has said, "When my code must fit into old and unstructured programs, then by necessity, my code also looks old and unstructured."

Another organization applies a policy which allows maintenance programmers to make the enhancement decision through one of two possible channels. First, if the individual feels after analyzing the change that a rewrite will be faster, then he or she may simply perform a rewrite of a bounded section. However, the individual

PROBLEM AREAS	ALL INTERVIEWED SUBJECTS	EMPLOYEES	MANAGERS
Poor Doc.	29.039	29.206	29.200
Poor Design	30.757	33.016	27.200
Poor Code	18.913	19.206	18.450
Time Frame	11.019	8.889	14.375
Other	11.175	9.683	13.525

FIXIT TASKS (% OF PROBLEMS)

PROBLEM AREAS	ALL INTERVIEWED SUBJECTS	EMPLOYEES	MANAGERS
Poor Doc.	25.931	28.387	22.125
Poor Design	30.971	31.516	30.125
Poor Code	17.265	17.194	17.375
Time Frame	14.363	13.548	15.625
Other	11.471	9.355	14.750

ENHANCEMENT TASKS (% OF PROBLEMS)

Perceived Problem Areas of Maintenance Efforts
By Interviewed Subjects

accepts responsibility for adhering to the original schedule and is also expected to meet requirements for documentation and testing of the section.

The other channel through which rewrites occur requires approval from the maintenance person's manager. In this case, the organization encourages individuals, after analyzing the requested change and the affected code, to prepare a brief request for time and resources to rebuild a bounded section of the system. Possible justifications include the need for general logic, very high start-up costs in understanding the module(s), failure histories, or frequent change histories. Responses to the requests may result in complete approval, an assignment to formally document the section, or a decision to delay the enhancement.

This organization has found that its maintenance personnel have responded positively to this policy. From a job design perspective, the reasons are clear. The two channels through which rewrites can occur give the individual maximum autonomy and simultaneously increase skill variety and task significance. This increase in the scope of the maintenance job does not reduce managerial control, nor does it force the shop into an environment in which it must rewrite all of its code. The result of the policy has been an observed increase in productivity, not only due to increased employee satisfaction due to a match of GNS and MPS, but also due to a reduction in task time for new changes requested on the updated code.

Summary: Development vs. Maintenance

This case has indicated that maintenance productivity is a product of job design and the policies of the organization towards the performance of maintenance and development. The acceptance of incompletely developed systems into maintenance obviously creates problems which go far beyond factors which can be controlled by the maintenance management. However, the documentation problem, and to a greater extent, the rewrite problem, can (and must) be addressed within the maintenance group.

Productivity of the maintenance activity is dependent not only on human behavior due to job design, but also on the nature of the task itself. The figures on the previous page, from this research, show clear agreement between management and professionals on factors which increase the difficulty, and lower the productivity, of maintenance tasks. For fixit and enhancement tasks, both management and employees ascribe over 70% of all difficulties to poor documentation, poor design, and poor code. As this material has shown, the maintenance group can address these issues and improve productivity two ways--through increases in employee motivation and decreases in problems associated with the activity.

APPLICATION CASE N
THE MAINTENANCE ROLE OF THE SENIOR PROFESSIONAL

John Enright sighed as he watched the office door close behind the departing Ron Jacobson. He had been enjoying the rounds of employee performance reviews until today, but this session with Ron had been frustrating and discouraging. He had been forced to explain to Ron that his subordinates on the remote order processing system project were complaining that he was not holding up his end of the development effort. Specifically, other employees were commenting on Ron's failure to respond to the urgency of the problematic development task. One person claimed that Ron was ". . . just riding with the development group without really helping out." Another team member said, "As much as I like and respect Ron, he just isn't putting as much into the project as he used to. With all that's riding on our success or failure, we can't afford to have members who give less than 100%." Finally, and most suprising, considering Ron's history of technical competence, a junior team member had complained that the new design did not reflect current technology.

Before the review, John had studied Jacobson's file in detail. Ron had started with the organization almost 19 years previously in the manufacturing side of the house. As years had gone by, he had been promoted and moved through the company until, due to the computerization of his current work area, he became involved with the systems group. He began in data entry, moved to operations, and from there gained entry into the programming staff.

Though Ron's performance reports had been quite satisfactory through his progression, they were particularly positive after the move into programming - almost as though he had finally found his niche. His move to programmer/analyst was rapid and smooth, as was his promotion to analyst. During that period, he was active in the development of literally every major system which was implemented in the organization. After his promotion to senior analyst and subsequent assignment as a project leader, his efforts were instrumental in the development of some of the organization's most critical systems, all of which were still in operation.

The only weak review in Ron's file occurred about six months after his assignment to the position of programming manager. In that review, he had been informed that his managerial skills simply were not sufficient and that it was deemed best that he move back to the project leader position where his skills could be best used.

Ron's manager at the time had added a personal note to that report indicating that the organization may have done him a disservice in promoting him in the first place. He said, "Ron was not promoted due to his own request. In fact, he stated at the time of the promotion that he wasn't sure that he wanted to be a manager and that he really might rather stay in the active development ranks."

After the move back into the senior analyst position, Ron Jacobson had again been every manager's dream come true. He had exerted herculean efforts to implement difficult systems and had been the foundation of knowledge when existing systems failed or required difficult modifications. His efforts as team leader continued to be satisfactory. In one review, his manager had written, "Ron is good at project prioritization, personnel assignment, and close supervision of immediate subordinates. As long as budgeting and other higher level administrative duties are not required of him, he should continue to serve an important role in his present position."

But now Ron's continuing role was in question. His previous review had a note about possible decreases in enthusiasm and commitment, and now Enright would be forced to make an even more negative set of entries into the file. In response to questions about his performance, Ron had freely admitted that his commitment was not what it had been in the past.

"I'm sorry," he said. "I understand your comments and I believe that they are fair, but I just can't make myself do it all anymore. I have fifteen years now in systems, and I'm just plain tired. I don't want to work nights and weekends anymore, and I no longer enjoy the challenge of committing to projects with high risks of failure. In the past, I have enjoyed the process of making projects work in spite of difficulties, but it's no fun anymore."

"No," he added. "Its worse than no fun. I really don't like the pressure. I've realized this for some time, but I've tried not to admit it to anyone. I have only a few years to my retirement and I can't afford to risk that. But since we're discussing apparent criticism from the rest of the staff, I guess I have to admit that I just can't make myself give that much anymore."

John Enright was new to the organization, having moved there into his first major management job only five months earlier, but he was familiar with problems of this type. In his old company, he had watched other senior employees lose the motivation to tackle complex new systems and fall out of the front lines of technology. Neither management or the employees had been able to find a satisfactory solution to the problem. In one case, the final resolution had been forced due to the employee's inability to work on the new systems. As the last of the old systems was phased out, so was the individual. It was not a situation which anyone at the company was pleased with, but no other avenues had been apparent. John had promised himself that he would do everything in his power to handle similar situations better, though he had not expected the occasion to arise so quickly.

That night, at home, John re-read the Couger and Zawacki book on the motivation of DP personnel in the hope of finding something which could shed some light on his problem. "What is it that is reducing Ron's apparent motivation," he mused, "and what responses are possible to preserve this valuable resource to the company and, at the same time, allow him to serve his remaining time to retirement with honor?"

Finally, John noted the documented evidence that GNS (growth need strength) is negatively correlated with age, and the picture began to clear. Persons with high-GNS seek jobs with high motivating potential, and a failure to achieve such a job results in low motivation and low productivity. He had known that relationship and applied it successfully in order to keep his employees motivated, but he had failed to consider the converse. "If high-GNS people are unsatisfied with low motivating jobs," he argued, "then a decrease in the GNS for a person in a highly motivating job must create a similar imbalance. If that is true, then my job becomes one of aiding Ron in finding a new match-up between GNS and MPS, just as I seek to do for my younger employees with the opposite problem."

The next day, John called Ron back to his office. He explained his thoughts on the issues, being careful to clearly indicate that a reduction in GNS was not any reflection on the individual, but an apparent natural occurrence as we grow older. "Given my thoughts," he said to Ron, "I would like for you and I to work together to seek a productive and satisfying assignment for you."

Continued discussions resulted in a mutual agreement that the single most valuable resource which Ron could offer to the organization was his vast knowledge of existing systems. In addition, they agreed that knowledge of that type was less time-sensitive than in earlier years due to the rapidly increasing average life of installed systems. Finally, they agreed that Ron's experience as a practicing professional, along with his group leadership abilities were also of value.

With this information, John talked with other managers, seeking to combine Ron's strengths in a position of mutual benefit to him and the organization. The solution which emerged was both simple and effective. John then called Ron back to his office for another discussion.

"Ron," he said. "We would like for you to continue functioning as a group leader, but in a slightly different capacity. We would like for you to accept maintenance responsibility for two of our largest and most important systems - systems which we expect to continue functioning for a number of years without redevelopment. We will provide a group of junior personnel who can handle small problems alone and implement your solutions to the more complex changes. We will be moving people in and out of that group as they gain sufficient skills to require more demanding jobs, but we promise to support you with an adequate staff at all times. While you will have final responsibility for the system, you will not have to do all of the work. We would hope that you will share your years of experience with these employees in order to help them advance in the systems field more rapidly."

After further discussions, Ron left the office again, but this time with a more positive stride than he had used around the office for a long time. It was clear that he was happy with the resolution and anxious to begin meeting his new responsibilities.

Discussion Of The Solution

The key to the solution to this problem lay in the recognition that the job characteristics which motivated Ron in the past had, in many cases, become demotivating factors due to a lessening growth need strength. The resulting mismatch between GNS and MPS created an unstable environment which, if uncorrected, would have made a continued utilization of Ron impossible. The effort to find a new assignment with the proper characteristics signaled the beginning of the solution.

The new job held a number of strong characteristics for Ron. Task significance was high, because the systems to be maintained were critical to the organization. In addition, the assignment of full responsibility for the effort resulted in high autonomy. However, skill variety dropped significantly from the massive development efforts. The availability of a support staff provided the balance needed in order to buffer Ron from most of the highly urgent efforts.

The issue of the older employee is seldom discussed in the systems industry, partly because they are an absolute minority and partly because we have really had no solutions to problems of the type discussed here. However, concepts of motivation and job design now allow us to actively match these valuable employees to jobs just as we do the same for their younger counterparts.

SECTION 4:
AGGREGATE RESEARCH RESULTS

CHAPTER 6
ANALYSIS OF MAINTENANCE DATA

In the study of maintenance activities, it is important to realize that it is impossible to control all other environmental variables in order to isolate the effects of the primary study factor. In this case, the problem is particularly severe because so many factors affect the primary measures of data processing employees (Couger and Zawacki, 1980). In order to clarify the effects of maintenance, this chapter presents a series of analyses which compare the study data with other data bases on various levels in order to establish a framework for discussion. This material is broken into a set of chapter sections which present the baseline research results in a series of discussions of increasing detail.

First, the entire set of research data is presented in a macro analysis in order to establish a set of important relationships. Here, the set of all respondents from all of the firms involved in the maintenance research is compared to the national data base which has been presented in the Couger and Zawacki book (1980). In addition, the set of interviewed employees is compared to the national data base and to the set of all respondents. The analysis of this chapter explains similarities and differences between these three data bases.

Next, the demographics of the interviewed subjects are discussed in detail. This group comprises the individuals from whom the most detailed information about employee perceptions of maintenance was obtained. In addition, this group represents the types of individuals who are being assigned to the maintenance effort by today's managers.

Finally, a detailed comparison of the interviewed personnel to all other respondents within each firm is presented. This supports discussions of differences between high and low maintenance personnel within specific organizations.

In Chapter 7, this discussion of research results continues. Based on the background analysis of Chapter 6, a series of relationships between levels of maintenance and the core job dimensions are discussed. In addition, the results of the structured interview questionnaire are presented.

MACRO ANALYSIS OF C-Z DATA

The national norms from the Couger and Zawacki book are the natural baseline for evaluation of the maintenance data. While the current size and scope of their data base is far larger than that reported in the book, those in the book have received the most public attention. They represent the only published statistical analysis of data processing employees' perceptions of their jobs and therefore are the basis of our comparisons.

Tables 6.1 and 6.2 show a direct comparison of the set of all respondents in this study to the Couger-Zawacki

VARIABLE	NATIONAL DATA		ALL RESPONDENTS		STAT TEST RESULTS
	MEAN (M1)	STD.DEV. (V1)	MEAN (M2)	STD.DEV. (V2)	
SEX	1.262	.450	1.342	.479	M2 GT M1 (.01)
AGE	2.939	.913	2.813	.945	M2 LT M1 (.01)
EDUCATION	4.594	.702	4.632	1.134	V2 GT V1 (.01)
MARITAL STATUS	1.626	.487	1.591	.544	
YEARS WITH FIRM	2.728	1.419	2.804	1.695	V2 GT V1 (.01)

For National Data, N = 709
For Maintenance Data, N = 555

NOTE...Significance levels are in parentheses in Results Column.

Variables are scored as follows:
- SEX : Male = 1; Female = 2.
- AGE : Under 20 = 1; Twenties = 2; Thirties = 3; Forties = 4; Fifties = 5; Sixties = 6.
- EDUC: Grade = 1; Some High = 2; High = 3; Some College = 4; B.A. or B.S. = 5; M.A. or M.S. = 6; Ph.D. = 7.
- MSTAT: Single = 1; Married = 2.
- YEARS: 1 or less = 1; 1 to 4 = 2; 4 to 8 = 3; 8 to 12 = 4; 12 to 16 = 5; Over 16 = 6.

PROGRAMMER AND ANALYST DEMOGRAPHICS
National Data Base vs. All Respondents
Table 6.1

national norms. The first table depicts the demographic data, and the second shows the job related variables from the research. The information in the tables includes the average scores and a measure of their variation from the relevant data sets and the results of a statistical analysis of the results. For each variable of interest, a set of hypotheses on the equality of the means and variances between the two groups was tested. The results of those statistical tests are also presented in the table. The numbers in parentheses indicate the level at which the results are significant. A significance level, for instance, of .01, indicates that the observed differences between the pair of means or variances is sufficiently large to indicate that there is a 99% or larger probability that the difference is not due to chance alone.

In Table 6.1, for instance, the means and standard deviations of the demographic variables of sex, age, education, marital status, and years with the firm are analyzed. Here, the two groups being compared are the national data base and the set of all respondents from all of the firms in the research. Note that this analysis shows that the average number of women data processing professionals is greater in this set of data. This result is not unexpected, given the increasing number of professional women in general. In addition, the average age of employees in the maintenance research group is significantly lower than that of the national norms. This reflects the influx of new, young professionals into the data processing field in response to the need for additional personnel since the original Couger and Zawacki work. Finally, the variations in education and years with the firm are greater than in the original data base. This is a natural result of a maturing industry.

VARIABLE	NATIONAL DATA		ALL RESPONDENTS		STAT TEST RESULTS
	MEAN (M1)	STD.DEV. (V1)	MEAN (M2)	STD.DEV. (V2)	
SKILL VARIETY	5.408	1.106	5.525	.983	M2 GT M1 (.05)
TASK IDENTITY	5.205	1.110	5.102	1.150	
TASK SIGNIFICANCE	5.605	1.176	5.486	1.146	M2 LT M1 (.05)
AUTONOMY	5.290	1.098	5.189	1.072	
FEEDBACK FROM JOB	5.133	1.066	5.005	1.104	M2 LT M1 (.05)
FDBK FROM SUPERVISORS	3.966	1.559	4.184	1.549	M2 GT M1 (.01)
GROWTH NEED STRENGTH	5.905	.994	6.217	.771	M2 GT M1 (.01) V2 LT V1 (.01)
MPS	153.581	62.464	146.609	64.292	M2 LT M1 (.05)

For National Data, N = 709
For Maintenance Data, N = 555

NOTE...Significance levels are in parentheses in Results Column.

All variables except MPS are scored on a 1 to 7 scale.

PROGRAMMER AND ANALYST JOB VARIABLES
National Data Base vs. All Respondents
Table 6.2

Many of the older professionals did not receive formal education in systems, but, rather, moved up through the ranks.

In Table 6.2, the same two groups are compared with respect to the core job dimensions of skill variety, task identity, task significance, autonomy, and feedback from the job. In addition, feedback from supervisors, growth need strength, and motivating potential score (MPS) are shown. Here we see that, compared to the national norms, the research group showed a generally high level of skill variety, though only at the .05 level. Task identity was slightly, but not significantly lower, and task significance was lower in the study firms at the .05 level. In addition, feedback from the job was significantly lower, while feedback from supervisors was higher. Growth need strength of the individuals in the study was actually higher than the national norm, though there was greater variation in this measure than in the national data. Finally, the motivating potential score, a summary measure of the overall richness of the job, was significantly lower than the national norms at the .05 level.

These results must be interpreted very carefully. The study firms for this project were carefully selected to represent healthy organizations with good reputations for development and maintenance. As such, they are not necessarily representative of the industry as a whole. Their results are therefore representative only of the study group and do not indicate industry norms. In addition, these firms, as examples of successful

VARIABLE	NATIONAL DATA		INTERVIEW DATA		STAT TEST RESULTS
	MEAN (M1)	STD.DEV. (V1)	MEAN (M2)	STD.DEV. (V2)	
SEX	1.262	.450	1.321	.471	
AGE	2.939	.913	3.018	1.000	
EDUCATION	4.594	.702	4.607	.802	
MARITAL STATUS	1.626	.487	1.714	.456	
YEARS WITH FIRM	2.728	1.419	3.143	1.710	M2 GT M1 (.05) V2 GT V1 (.05)

For National Data, N = 709
For Maintenance Data, N = 56

NOTE...Significance levels are in parentheses in Results Column.

Variables are scored as follows:
SEX : Male = 1; Female = 2.
AGE : Under 20 = 1; Twenties = 2; Thirties = 3; Forties = 4; Fifties = 5; Sixties = 6.
EDUC: Grade = 1; Some High = 2; High = 3; Some College = 4; B.A. or B.S. = 5; M.A. or M.S. = 6; Ph.D. = 7.
MSTAT: Single = 1; Married = 2.
YEARS: 1 or less = 1; 1 to 4 = 2; 4 to 8 = 3; 8 to 12 = 4; 12 to 16 = 5; Over 16 = 6.

PROGRAMMER AND ANALYST DEMOGRAPHICS
National Data vs. All Interviewed Subjects From All Study Firms

Table 6.3

organizations, may also exhibit the effects of improved systems knowledge, management expertise, and employee characteristics.

For instance, the higher level of feedback from supervisors may result from a combination of these factors. While the study firms have good reputations for being well managed, one would expect that the industry as a whole is probably benefiting from more experienced and better trained managers. The relative effects of these factors cannot be completely clarified in this work. The differences have been noted simply in order to provide a baseline for the comparison of high and low maintenance personnel within the study organizations.

Table 6.3, above, shows the results of a comparison of the set of all interviewed subjects from all of the firms in the study to the national norms from Couger and Zawacki. As discussed earlier in this book, the interviewed personnel included management and professionals. Management was interviewed in a group, and professionals were involved in individual discussions of at least one hour. In order to provide a valid comparison to the national norms on programmers and analysts, only the data from the professionals is included in this analysis. Responses from management are discussed in Chapter 7.

It is important to remember that this set of professionals was <u>not</u> chosen randomly from the study organizations. In each organization, management was asked to identify a set of programmers and analysts who they perceived to be heavily involved in maintenance. The following discussion serves to place the resulting interviewed group in

VARIABLE	NATIONAL DATA		INTERVIEW DATA		STAT TEST RESULTS
	MEAN (M1)	STD.DEV. (V1)	MEAN (M2)	STD.DEV. (V2)	
SKILL VARIETY	5.408	1.106	5.59	.84	V2 LT V1 (.05)
TASK IDENTITY	5.205	1.110	5.28	1.05	
TASK SIGNIFICANCE	5.605	1.176	5.71	1.22	
AUTONOMY	5.290	1.098	5.39	.86	V2 LT V1 (.01)
FEEDBACK FROM JOB	5.133	1.066	5.15	1.02	
FDBK FROM SUPERVISORS	3.966	1.559	4.49	1.32	M2 GT M1 (.01)
GROWTH NEED STRENGTH	5.905	.994	6.18	.69	M2 GT M1 (.01) V2 LT V1 (.01)
MPS	153.581	62.464	153.42	64.29	

For National Data, N = 709
For Maintenance Data, N = 55

NOTE...Significance levels are in parentheses in Results Column.

All variables except MPS are scored on a 1 to 7 scale.

PROGRAMMER AND ANALYST JOB VARIABLES
National Data vs. All Interviewed Subjects From All Study Firms
Table 6.4

perspective with respect to the national norms and in comparison to the set of all responding professionals from each of the study organizations.

From Table 6.3 it is clear that very little significant difference exists between the interviewed subjects and the national data base. The only notable finding here involves the significantly higher number of years with the firm for the interviewed employees. This implies that maintenance personnel may be drawn from the pool of senior employees, rather than from the ranks of new employees.

This finding must be carefully considered in order to avoid inappropriate conclusions. First, a few of the study organizations did, indeed, assign very senior employees to the maintenance task. (Case N, in Chapter 5, discusses one reason for this type of assignment.) Given the average age of professionals in the data processing industry (generally in the late twenties), a few employees close to the retirement age can significantly affect the average for a group of this size. Indeed, Table 6.3 does show a significantly high variation in age for this group, indicating that one should not assume that maintenance is being performed in general by senior employees. Rather, these findings reflect the fact that organizations may choose to utilize junior or senior employees with positive results, depending on the motivational nature of the resulting assignments.

In Table 6.4, above, the job related variables for the interviewed group are compared to the C-Z national norms. Here we find significantly higher feedback from supervisors and growth need strength. Again, these

VARIABLE	ALL RESPONDENTS		INTERVIEW DATA		STAT TEST RESULTS
	MEAN (M1)	STD.DEV. (V1)	MEAN (M2)	STD.DEV. (V2)	
SEX	1.342	.479	1.321	.471	
AGE	2.813	.945	3.018	1.000	
EDUCATION	4.632	1.134	4.607	.802	V2 LT V1 (.01)
MARITAL STATUS	1.591	.544	1.714	.456	M2 GT M1 (.05)
YEARS WITH FIRM	2.804	1.695	3.143	1.710	

For All Respondents, N = 555
For Maintenance Data, N = 56

NOTE...Significance levels are in parentheses in Results Column.

Variables are scored as follows:
- SEX : Male = 1; Female = 2.
- AGE : Under 20 = 1; Twenties = 2; Thirties = 3; Forties = 4; Fifties = 5; Sixties = 6.
- EDUC: Grade = 1; Some High = 2; High = 3; Some College = 4; B.A. or B.S. = 5; M.A. or M.S. = 6; Ph.D. = 7.
- MSTAT: Single = 1; Married = 2.
- YEARS: 1 or less = 1; 1 to 4 = 2; 4 to 8 = 3; 8 to 12 = 4; 12 to 16 = 5; Over 16 = 6.

PROGRAMMER AND ANALYST DEMOGRAPHICS
All Respondents vs. All Interviewed Subjects From All Study Firms
Table 6.5

results must be interpreted carefully. While the averages are, indeed, higher than the national norms, Table 6.2 indicated that the study group as a whole had relatively high scores for these variables, also. It is clear that this result may be due to the overall characteristics of the study group, and not to any characteristics of high maintenance personnel.

The most meaningful results in Table 6.4 are indicated in the significance of variation in scores on skill variety and autonomy. Individuals involved in high maintenance tasks apparently have similar perceptions about certain aspects of those tasks, resulting in low variation in their scores on these job variables.

In order to complete the clarification of the characteristics of the interviewed group of employees, Tables 6.5 and 6.6 compare that group to the set of all respondents in all of the study firms. This comparison serves to identify any significant differences between the high maintenance group and the entire set of professionals from their organizations.

Table 6.5 shows that a higher than average number of the selected employees were married, and that there was significantly less variation in the amount of education held by that group. Due to the selection process, it is difficult to determine any major underlying issue relating to these differences. They are simply noted in the baseline analysis.

In Table 6.6, on the following page, the interview group shows very few significant differences from the overall study group when compared with respect to the job

VARIABLE	ALL RESPONDENTS		INTERVIEW DATA		STAT TEST RESULTS
	MEAN (M1)	STD.DEV. (V1)	MEAN (M2)	STD.DEV. (V2)	
SKILL VARIETY	5.525	.983	5.59	.84	
TASK IDENTITY	5.102	1.150	5.28	1.05	
TASK SIGNIFICANCE	5.486	1.146	5.71	1.22	
AUTONOMY	5.189	1.072	5.39	.86	V2 LT V1 (.05)
FEEDBACK FROM JOB	5.005	1.104	5.15	1.02	
FDBK FROM SUPERVISORS	4.184	1.549	4.49	1.32	
GROWTH NEED STRENGTH	6.217	.771	6.18	.69	
MPS	146.609	64.292	153.42	64.29	

For All Respondents, N = 555
For Maintenance Data, N = 55

NOTE...Significance levels are in parentheses in Results Column.

All variables except MPS are scored on a 1 to 7 scale.

PROGRAMMER AND ANALYST JOB VARIABLES
All Respondents vs. All Interviewed Subjects From All Study Firms
Table 6.6

variables. The only significant result in this table involves autonomy, where the interviewed group showed a lower variation in their scores.

The implications in these results are clear. The set of all employees involved in the study varied in certain ways from the national norms established by Couger and Zawacki. However, the interviewed group appears to be very similar to the larger group of employees from which it was chosen. The only real trend seen in the interviewed group involves a set of lower variances compared to both the national norms <u>and</u> the study group as a whole. The indication is that these employees are similar to the rest of the professionals in the study, but that, due to the similarity of their work (maintenance), they show lower variation in their responses.

In addition, the lack of differences in the job variables between the interviewed personnel and their peers within the same organizations supports the argument that these organizations are, in general, handling maintenance well. However, these statistics can result in misinterpretation in this area. The key point is that, at this level, the statistics indicate that the entire set of study organizations, taken as a whole, shows little difference between high maintenance personnel and the rest of their professionals. When the organizations are studied individually, more significant differences are found. The wide variation in the manner in which each organization addressed its maintenance problems resulted in local variations in scores which did not show up in the overall analysis which has just been presented.

Firm	MALES		FEMALES	
	Number	Percent	Number	Percent
42	6	75.0	2	25.0
43	4	57.1	3	42.9
45	3	37.5	5	62.5
46	4	57.1	3	42.9
48	4	66.7	2	33.3
491	7	77.8	2	22.2
492	6	100.0	0	0.0
50	4	80.0	1	20.0
All Firms	38	67.9	18	32.1

DISTRIBUTION OF SEX VARIABLE FOR INTERVIEWED SUBJECTS
Table 6.7

In order to better clarify characteristics of the interview group, the following section presents a more detailed set of data on its demographics.

DEMOGRAPHICS OF INTERVIEWED SUBJECTS

In general, the demographics of the high maintenance personnel who were selected by their managers to participate in the interviews varied greatly between firms. As noted earlier, some firms utilized older professionals, while others assigned the maintenance efforts to primarily younger employees. The series of tables from 6.7 through 6.11 present the demographics of the interviewed subjects, broken down by organization in order to illustrate this variation. It is important to remember that, while each firm has a reputation for successful management of the maintenance function, they differed greatly in terms of organization and personnel selection. These differences support the argument that maintenance productivity is related to the match between individual personnel characteristics and job characteristics, and that the problem is not simply a personnel issue.

Table 6.7, above, illustrates the variation in the distribution of males and females in the interviewed group. Here, we see that one firm chose 100% of its interview candidates as males, and one organization selected 62% of its interviewees as female. In general, the study did not discover any relationships between the sex variable and perceptions of the maintenance task.

This table exhibits a characteristic which exists across all of the five tabulated data presentations. The number of interviewed personnel varies widely across the organizations, and the total number of individuals may vary from table to table. This is due to two primary factors. First, this data results from the Couger and Zawacki instrument, and certain individuals did not answer every question. Second, not all of the professionals identified by management were actually high maintenance employees. Data for these individuals were deleted from all of the analyses which involved high maintenance personnel.

In Table 6.8, the distribution of the age variable is presented. Here we see that the interviewed employees from Firm 45 and Firm 491 were all under 40, while the interviewees from Firm 50 were all between forty and fifty. While these two organizations exhibit extremes in terms of age distribution, the other firms are interesting, also. Firm 492 shows a clear bimodal assignment, with young personnel assisting older employees, while Firm 46 shows a suprisingly even distribution of age.

The findings on the distributions of age among the interviewed employees are similar to other findings on the next two pages. In all of the tables, it is clear that each of these organizations has developed its own criteria for the assignment of individuals to the maintenance task. There is little or no similarity between the groups of high maintenance professionals in different organizations. This clearly supports the concept that the effectiveness of the maintenance operation need not depend on the individual characteristics of the personnel assigned to the effort. Instead, it is the organization and management of the maintenance task which is the primary factor in motivation and productivity. The cases presented in Section III illustrate some of the many combinations of personnel and job characteristics which create a positive maintenance environment.

INTERVIEWED SUBJECTS BY FIRM

While the previous section discussed the demographics of the interviewed professionals, this section presents more

Firm	UNDER 20		20 TO 29		30 TO 39		40 TO 49		50 TO 59		OVER 59	
	#	%	#	%	#	%	#	%	#	%	#	%
42	0	0.0	4	50.0	2	25.0	2	25.0	0	0.0	0	0.0
43	0	0.0	0	0.0	5	71.4	2	28.6	0	0.0	0	0.0
45	0	0.0	5	62.5	3	37.5	0	0.0	0	0.0	0	0.0
46	0	0.0	3	42.9	2	28.6	1	14.3	1	14.3	0	0.0
48	0	0.0	1	16.7	2	33.3	3	50.0	0	0.0	0	0.0
491	0	0.0	7	77.8	2	22.2	0	0.0	0	0.0	0	0.0
492	0	0.0	2	33.3	1	16.7	0	0.0	3	50.0	0	0.0
50	0	0.0	0	0.0	0	0.0	5	100.0	0	0.0	0	0.0
All Firms	0	0.0	22	39.3	17	30.4	13	23.2	4	7.1	0	0.0

DISTRIBUTION OF AGE VARIABLE
FOR INTERVIEWED SUBJECTS
Table 6.8

Firm	H.S.		SOME COL.		BS / BA		MS / MA		Ph.D.	
	#	%	#	%	#	%	#	%	#	%
42	0	0.0	3	37.5	3	37.5	1	12.5	1	12.5
43	1	14.3	4	57.1	1	14.3	1	14.3	0	0.0
45	0	0.0	2	25.0	6	75.0	0	0.0	0	0.0
46	0	0.0	3	42.9	3	42.9	1	14.2	0	0.0
48	0	0.0	5	83.3	0	0.0	0	0.0	1	16.7
491	0	0.0	0	0.0	9	100.0	0	0.0	0	0.0
492	1	16.7	3	50.0	2	33.3	0	0.0	0	0.0
50	0	0.0	4	80.0	1	20.0	0	0.0	0	0.0
All Firms	2	3.6	24	42.9	25	44.7	3	5.4	2	3.6

DISTRIBUTION OF EDUCATION
FOR INTERVIEWED SUBJECTS
Table 6.9

detail on the job variable scores of that group, broken down by organization. Again, these data require careful consideration to avoid making unsupportable generalizations. First, remember that the group is purposely chosen to represent high maintenance individuals, and that it is not representative of the organizations individually or as a whole. In fact, as the previous discussion points out, interviewed personnel who did not, in fact, see themselves as having a maintenance role were not included in these analyses. Second, the sample sizes for the interviewed groups within each organization are too small to support statements about the significance of the results for an individual firm.

The purpose of the presentation of this set of figures is to illustrate the widely varying nature of interviewees' responses on the core job variables between organizations. These varying responses are primarily due to the organization of the maintenance effort for each of the firms. In addition, the amount of maintenance effort observed on individuals varied widely between the study organizations. As will be discussed in Chapter 7, many of the core job variables are directly correlated with the actual amount of maintenance being performed.

These figures, therefore, summarize job perceptions of high maintenance individuals for each of the study organizations. In each case, the values are compared to those of all of the study respondents within the organization in order to obtain a common comparison base.

Firm	SINGLE		MARRIED	
	Number	Percent	Number	Percent
42	1	12.5	7	87.5
43	2	28.6	5	71.4
45	3	37.5	5	62.5
46	1	14.3	6	85.7
48	2	33.3	4	66.7
491	3	33.3	6	66.7
492	3	50.0	3	50.0
50	1	20.0	4	80.0
All Firms	16	28.6	40	71.4

DISTRIBUTION OF MARITAL STATUS FOR INTERVIEWED SUBJECTS
Table 6.10

Firm	UNDER 1		1 TO 4		4 TO 8		8 TO 12		12 TO 16		OVER 16	
	#	%	#	%	#	%	#	%	#	%	#	%
42	0	0.0	6	75.0	2	25.0	0	0.0	0	0.0	0	0.0
43	0	0.0	2	28.6	2	28.6	0	0.0	1	14.3	2	28.6
45	0	0.0	4	50.0	1	12.5	1	12.5	1	12.5	1	12.5
46	0	0.0	5	71.4	2	28.6	0	0.0	0	0.0	0	0.0
48	0	0.0	3	50.0	3	50.0	0	0.0	0	0.0	0	0.0
491	3	33.3	4	44.4	1	11.1	0	0.0	1	11.1	0	0.0
492	2	33.3	0	0.0	0	0.0	0	0.0	0	0.0	4	66.7
50	0	0.0	0	0.0	0	0.0	0	0.0	1	20.0	4	80.0
All Firms	5	8.9	24	42.9	11	19.6	1	1.8	4	7.1	11	19.6

DISTRIBUTION OF YEARS WITH ORGANIZATION FOR INTERVIEWED SUBJECTS
Table 6.11

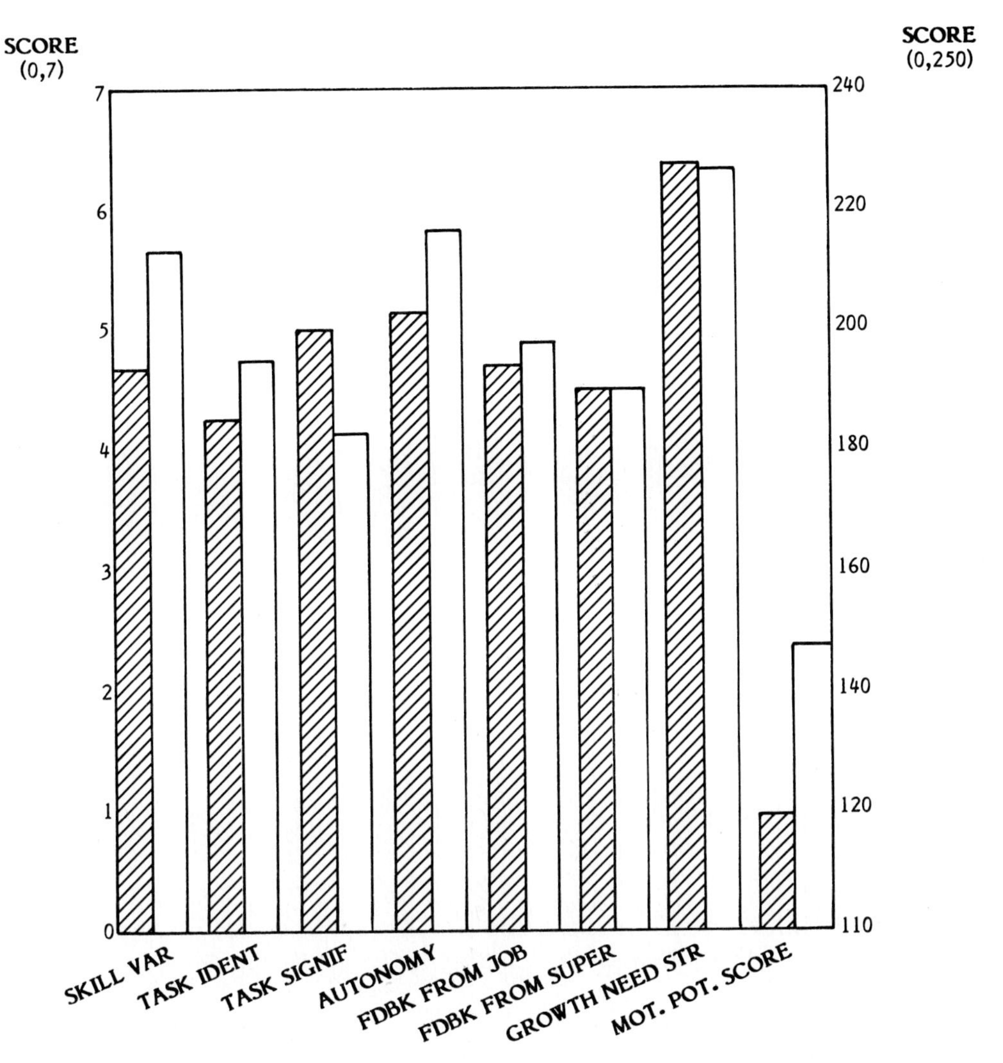

Firm 42: Respondents vs Interviewees
Figure 6.1

In Figure 6.1, above, the data for Firm 42 is presented. In this organization, almost all of the measures are at or well below those of the organization as a whole. In fact, the only measure which lies above the organization average is that of growth need strength. The high-GNS, paired with a very low motivating potential score, indicates a motivational mismatch. Indeed, this organization stood out in the study as having the most unhealthy maintenance structure of all of the involved firms.

In Firm 43, detailed in Figure 6.2, the core job variables are clustered around the group averages. While task significance was judged lower than average by the interviewed employees, all of the other scores were slightly above the average. The resulting MPS, matched with a high GNS, appears to be healthy for the organization.

By comparison, Firm 45 (Figure 6.3) shows a major lack of balance between MPS and GNS. While the GNS for the interviewed group is relatively high, the motivating potential score is low. Like Firm 42, this organization exhibited severe problems when viewed from the perspective of the high maintenance individuals. However, the organization, when analyzed from the perspective of all of its data processing professionals, exhibited even worse problems. In this case, the

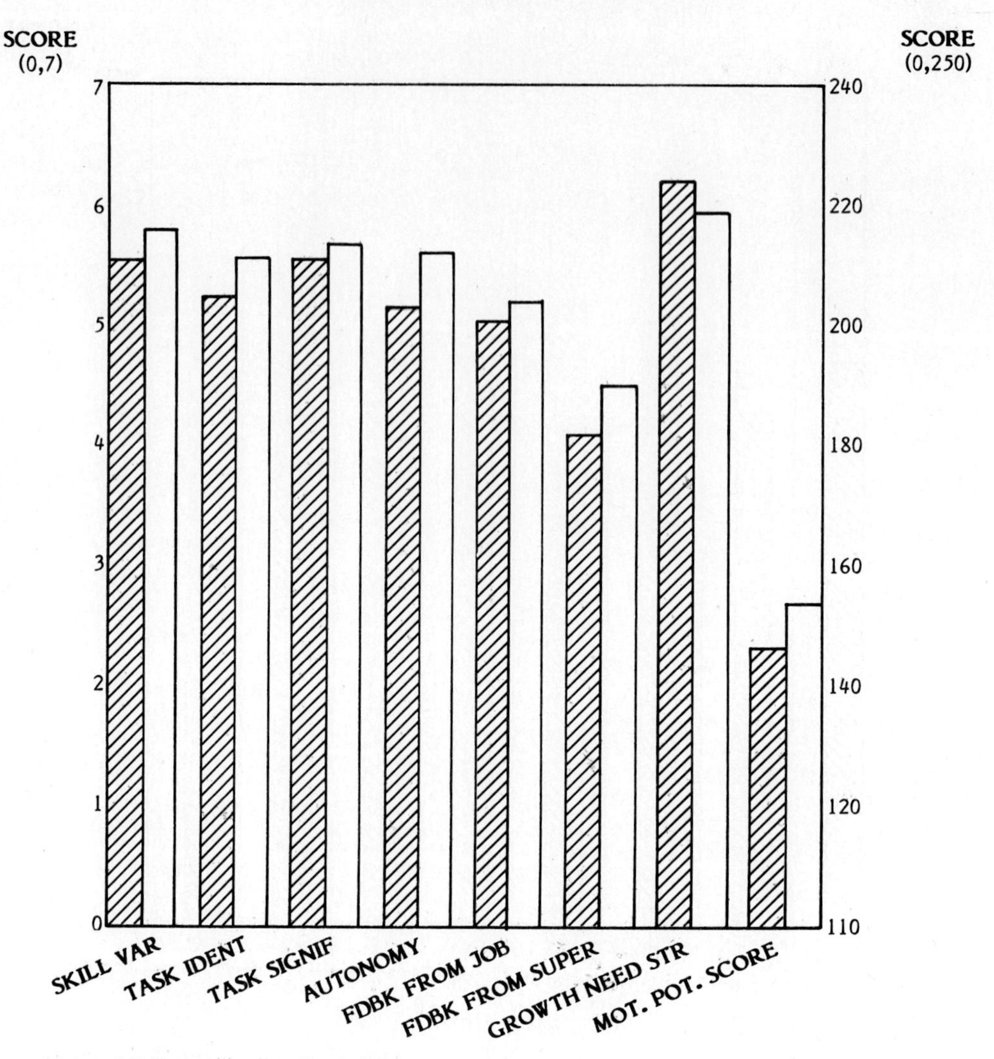

Firm 43: Respondents vs Interviewees
Figure 6.2

mismatch between GNS and MPS was reduced in the high maintenance employees.

The remaining figures, representing the interviewed employees of other organizations, are presented following this discussion. They are placed in this section of the report in order to provide further perspective on the professionals who participated in the interview process. In each case, those individuals' responses are compared to the group of all respondents in the study.

PRIMARY DATA ANALYSIS SUMMARY

This chapter has presented a set of preliminary findings based on the analysis of the maintenance data. The purpose of this set of discussions has been to provide a baseline for the detailed analyses which appear in the following chapter. There, the specific maintenance-related findings are discussed in detail.

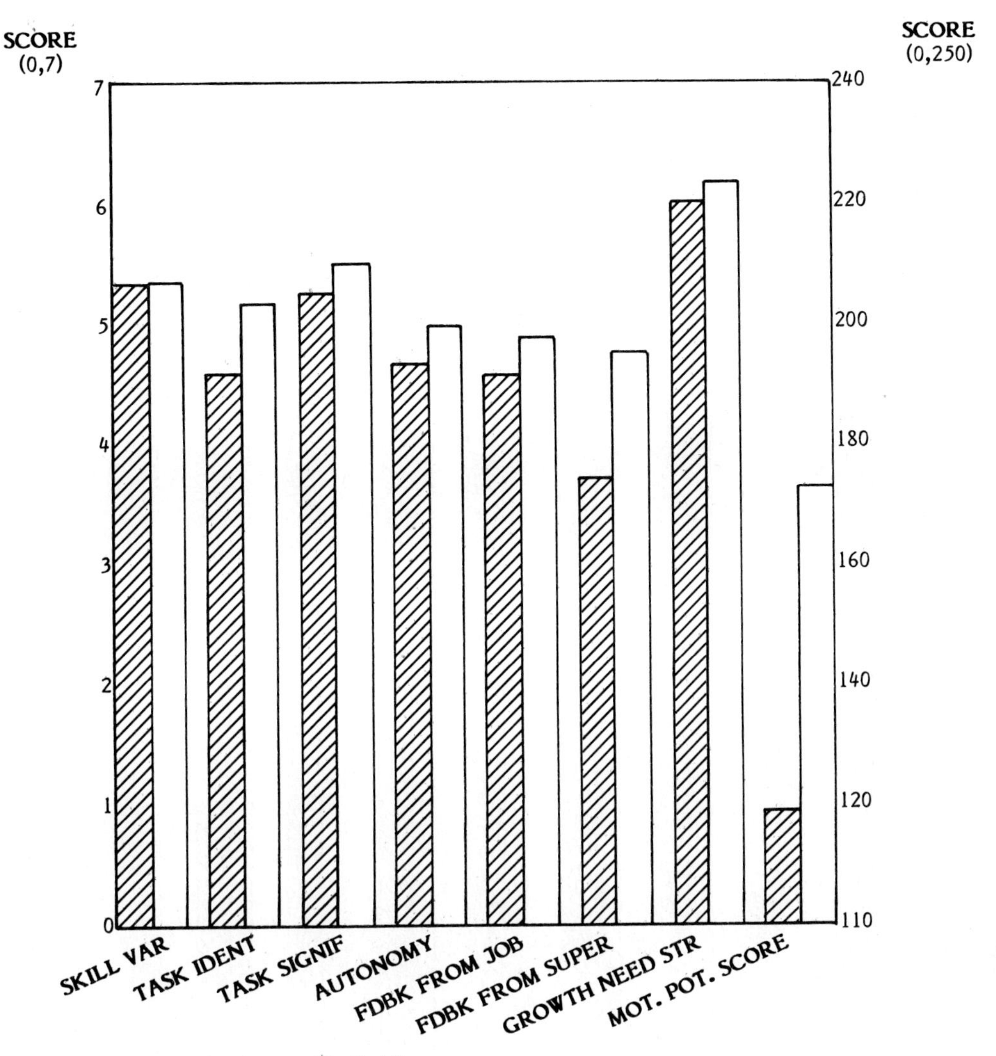

Firm 45: Respondents vs Interviewees
Figure 6.3

It is clear that, while some of the demographics of the interviewed personnel are significantly different from the national norms established by Couger and Zawacki, they are similar, in most respects, to other professionals within their own organizations. Therefore, it is safe to assume that the primary difference on job variables between the interview group and the rest of the respondents lies in their maintenance assignments. This difference is of primary interest in the study of the maintenance task.

The following chapter presents a series of discussions which treat various aspects of the maintenance effort. First, the relationship between the core job dimensions and the amount of maintenance assigned to an individual is explored. Next, the detailed results of the structured questionnaire are discussed in detail.

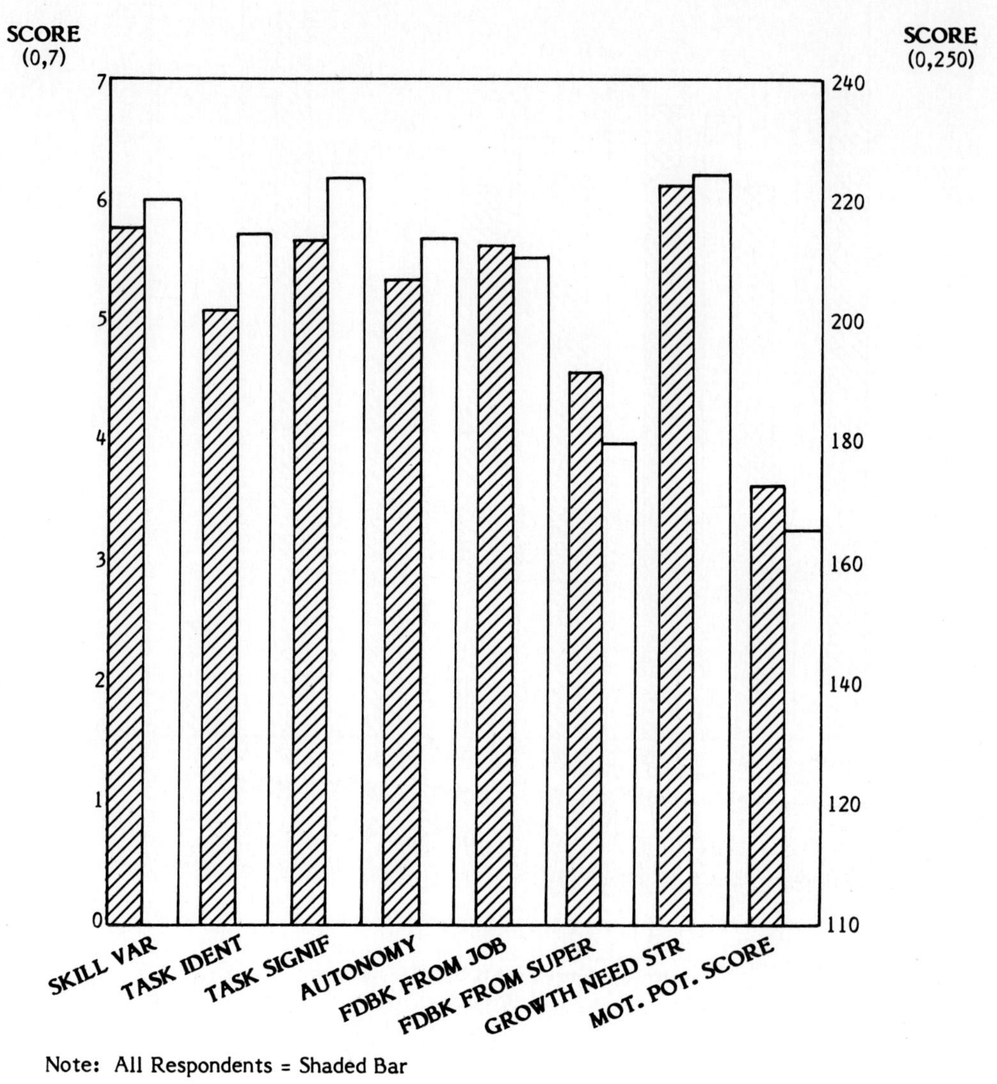

Note: All Respondents = Shaded Bar

Firm 46: Respondents vs Interviewees
Figure 6.4

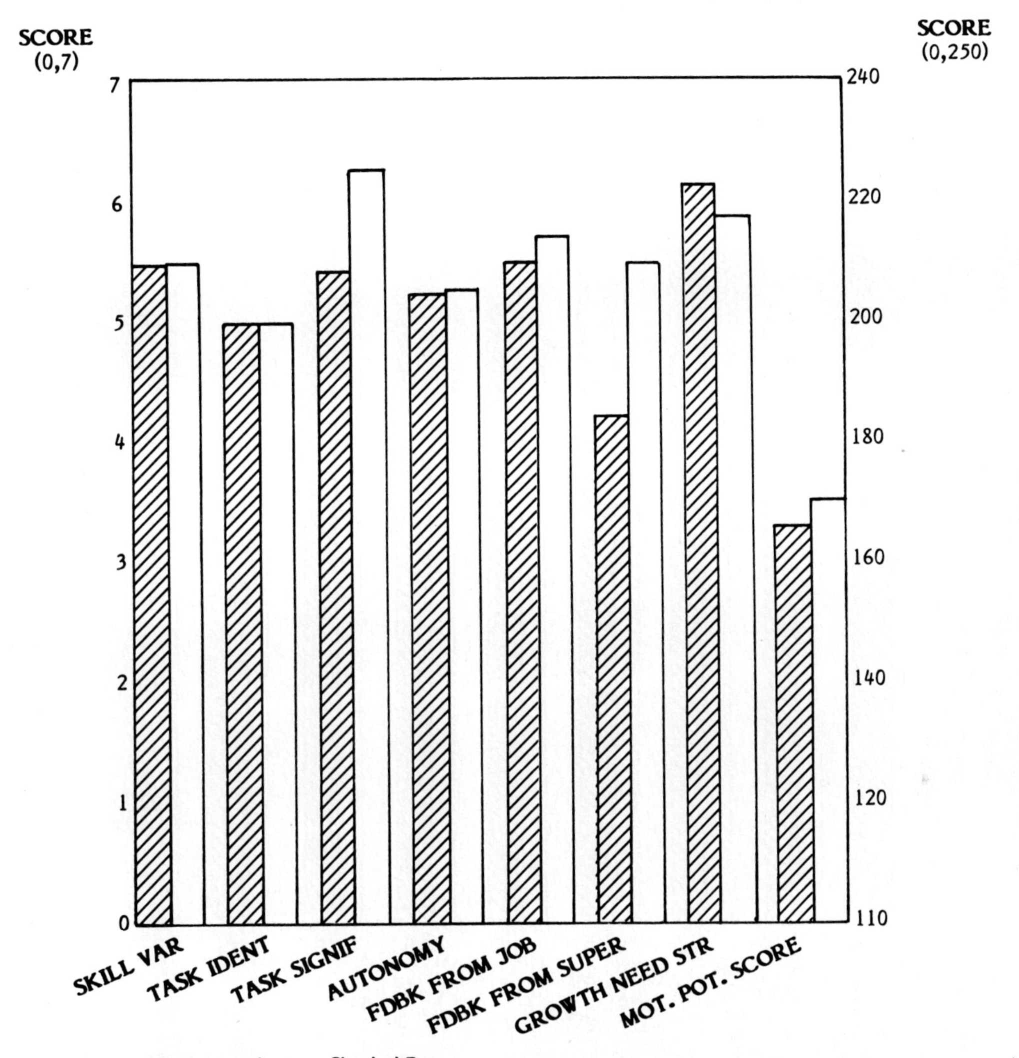

Note: All Respondents = Shaded Bar

Firm 48: Respondents vs Interviewees
Figure 6.5

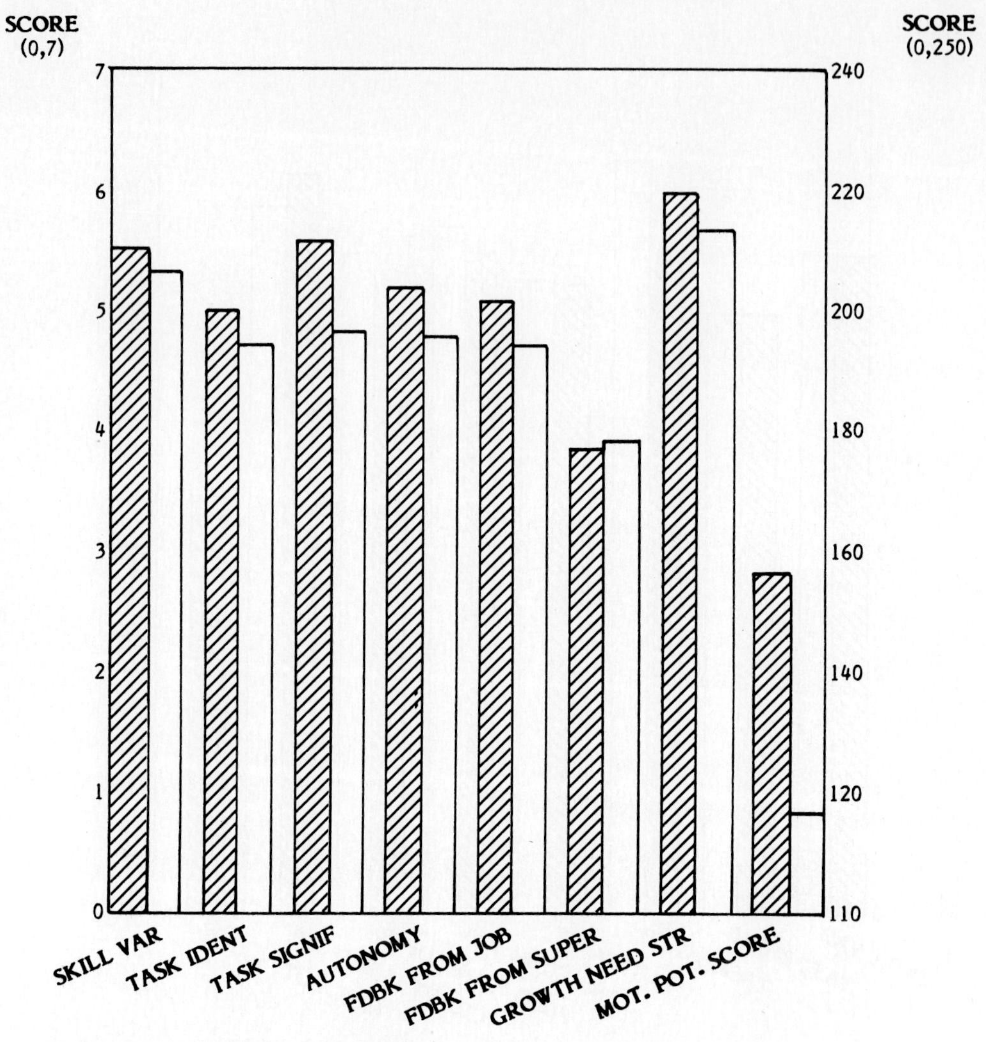

Note: All Respondents = Shaded Bar

Firm 50: Respondents vs Interviewees
Figure 6.6

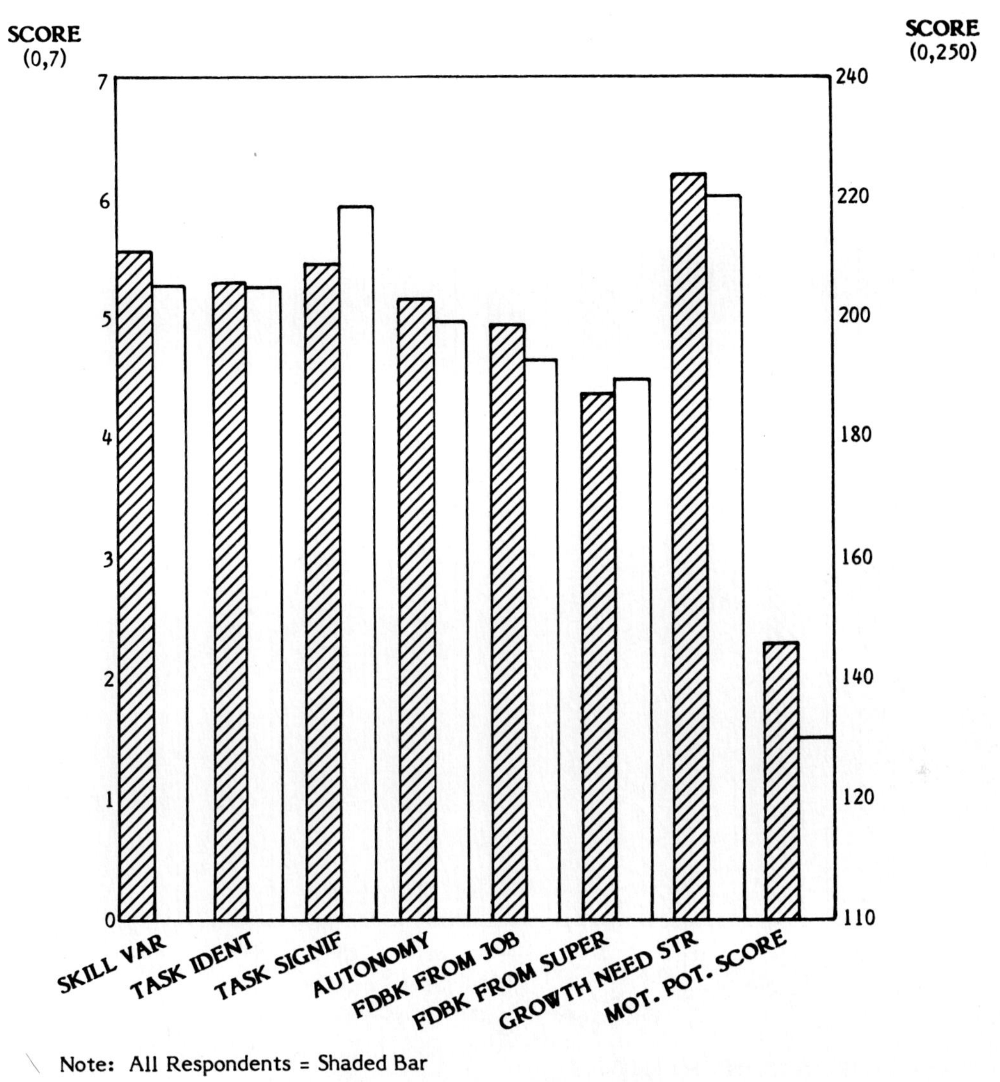

Note: All Respondents = Shaded Bar

Firm 491: Respondents vs Interviewees
Figure 6.7

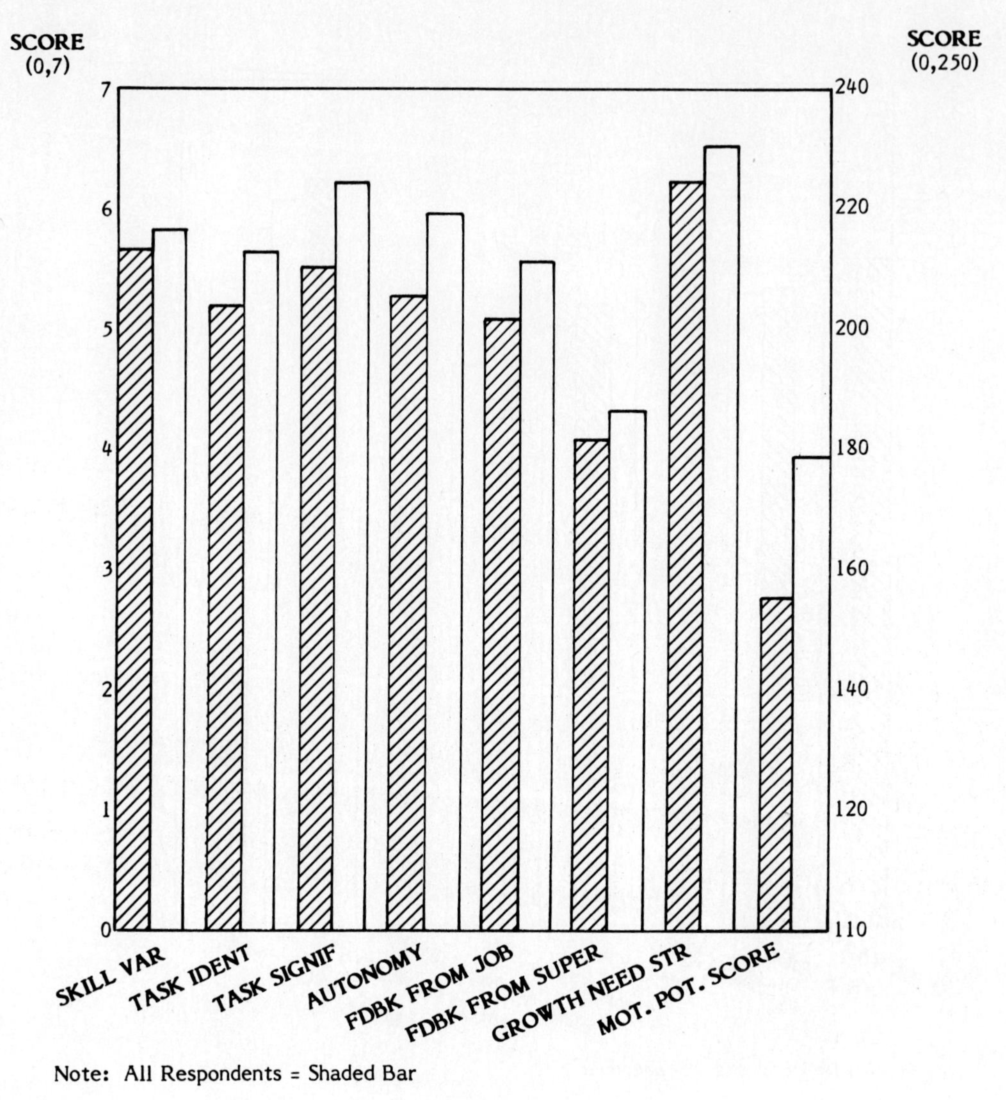

Note: All Respondents = Shaded Bar

Firm 492: Respondents vs Interviewees
Figure 6.8

CHAPTER 7
AGGREGATE RESULTS

While Chapter 6 presented a baseline analysis of the individuals involved in the research project, this chapter details the specific results of the maintenance-oriented portions of the study. Here, results from use of the Couger-Zawacki survey instrument and from the structured interview questionnaire are discussed in order to clarify the relationship between maintenance activities and individuals' job perceptions. In addition, comparisons are made between interviewees and their management with respect to their perceptions of aspects of the maintenance task, itself.

Before an individual completed the C-Z survey instrument, he or she was asked to delineate the percent of time spent in each of several classes of activity. These classes included analysis, design, new program development, programming, and maintenance.

It is important to note that no definition of maintenance was provided to respondents at this point. All responses were based on the respondent's personal definition of the term. The reason for this approach is that the definition of maintenance varies greatly throughout the literature and within the industry itself. By avoiding definitional problems, the instrument was able to elicit the amounts of maintenance which individuals perceived that they performed, irrespective of our official definition of the term. This in no way reduces the significance of the resulting data. The approach was consistent with industry studies designed to determine the percent of budget spent on new development versus maintenance. Definitions are rarely provided to respondents of those surveys. By allowing individual definitions of maintenance, the instrument was able to relate perceived maintenance responsibilities to perceptions about the resulting job. These results are presented first in this chapter.

During the interviews, each individual was again asked to estimate the percentage of his or her total work effort which involved maintenance activities. Again, this question was asked without providing any definition of the term. After the individual responded to the question, he or she was then asked to provide a definition of the effort to which the response referred. Contrary to much of the published literature on the subject of the definition of maintenance, the responses to this question were uniformly simple and straightforward.

The literature on the definition of maintenance contains complex definitional structures for this type of effort. For example, one such structure defines maintenance through a set of sub-components as:

- -Emergency Repairs
- -Corrective Coding
- -Upgrades
- -Changes in Conditions
- -Growth
- -Enhancements
- -Support.

Another definitional structure, with somewhat less complexity, is provided by Swanson[1] as:

-Corrective Maintenance
-Adaptive Maintenance
-Perfective Maintenance

While definitions of this type may be of value in the classification of specific maintenance efforts, this research found that they fail to parallel the manner in which individuals perceive the maintenance task. In every case, the interviewed personnel, through their own definitions, supported a binary classification of maintenance.

First, all of the individuals perceived a class of maintenance effort which is directed at existing systems because of failure. These efforts must be performed in order to make the system perform and they generally involve time pressures related to actual processing schedules. This class of effort is referenced in this study as being **Fixit** work. In other words, the system or problem must be fixed in order to support further operation.

The second class of activity discussed by these individuals involved all other changes to the system. While the literature discusses multiple sub-classifications for this type of effort, the individuals in the study clearly did not support that type of detailed break-down. They perceived that maintenance work was either necessary simply to make the system function (Fixit effort), or that it was required due to some external stimulus. The details of the source of the latter requirement were considered to be relevant only in terms of performing the work and meeting the requirement. Otherwise, all of these efforts were combined into a single class of maintenance which we refer to as **Enhancement.**

This classification of all maintenance effort into two types, fixit and enhancement efforts, was common to all of the individuals involved in the interview process. While the more detailed classifications may be of value in certain types of research, this research found that the personnel involved in the maintenance effort simply do not perceive the effort at that level of detail.

After the discussion of maintenance data from the Couger-Zawacki survey instrument, the discussion of the interview questionnaire will use the fixit/enhancement classification to provide further detail on the respondents' perceptions of the maintenance task.

MAINTENANCE: ALL RESPONDENTS

The average percent of maintenance activities reported by all programmers, programmer/analysts, and analysts, across all of the firms in the study, was 32%. Averages for individual firms varied from a low of 23% to a high of 45%. Within each of the firms, the standard devlation was large. This indicates that these organizations did not equally distribute maintenance across all of their employees. Instead, the amount of maintenance effort varied greatly for individuals.

In order to explore the relationship between the demographic and job variables and the amount of maintenance activities, the responses were categorized into a set of five classes. These classes grouped responses with 0 to 20%, 21 to 40%, 41 to 60%, 61 to 80%, and 81 to 100% of maintenance activities. In addition, a class of all respondents reporting 50% or more of their time in maintenance was analyzed.

The detailed data from these analyses is not presented in this chapter. It involves a large number of tables which are best summarized in the following figures and discussion. However, in order to provide the research detail, those tables are located in Appendix IV.

The first important issue for discussion relates to the nature of the findings presented in this section. Note carefully that these results compare all of the respondents from all of the study organizations, with respect to their maintenance responsibilities. This means that the number of individuals is large enough for statistical analysis, enabling statements about the significance of results. Because the numbers within a firm were too low to make statistical statements about relationships between the high maintenance interviewees and other respondents, this chapter provides much stronger results.

The demographics of the set of all respondents were first studied with respect to the amount of maintenance activity. For this set of data, significant differences

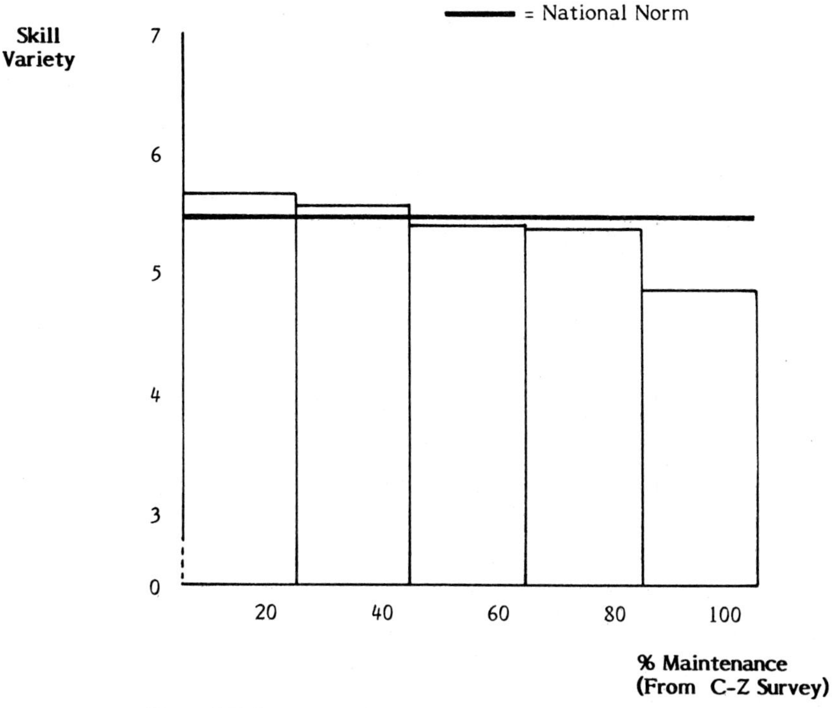

Note: All data was collected as interval data.

Figure 7.1: Skill Variety vs. % Maintenance (All Respondents)

were found between high and low maintenance personnel. At the .01 level of significance, it appears that high maintenance individuals are younger, better educated, and newer to the firm than those involved in low maintenance. Given the associated finding that education and years with the firm are strongly correlated with age, the important finding here is clear. Across all of the organizations studied, the highest levels of maintenance responsibilities tend to be assigned to the younger employees, supported with relatively few senior professionals.

When the relationships between job variables and maintenance responsibilities were studied, a set of important trends were shown to exist. Figure 7.1 exhibits such a relationship between skill variety and maintenance levels. Clearly, higher levels of maintenance efforts are negatively related to perceived skill variety. While very low levels of maintenance resulted in scores well above the national norms for skill variety, high maintenance responsibilities were tied to extremely low scores. The effect was most noticeable for the class of individuals with 80% or more of their time spent in maintenance.

In Figure 7.2, much the same trends occur when scores on task identity are studied. While the relationship between task identity and maintenance is not as clear as that for skill variety, its direction is still meaningful and significant.

This type of relationship existed for all of the core job variables, except for that of task significance (Figure 7.3). No significant correlation between maintenance and task significance was found. This result was clarified during the interviews, when it was found that task significance seems to be most related to other factors, such as the degree of isolation from users, rather than to the degree of maintenance activities.

Figure 7.6, showing the relationship between feedback from supervisors and maintenance, indicates that, for most of the study group, supervisor feedback is well above the national norms. It is not suprising that this group of healthy organizations shows high scores on this variable. Nevertheless, maintenance personnel in the over 80% category are significantly different. One of the observations of this study was that extremely high

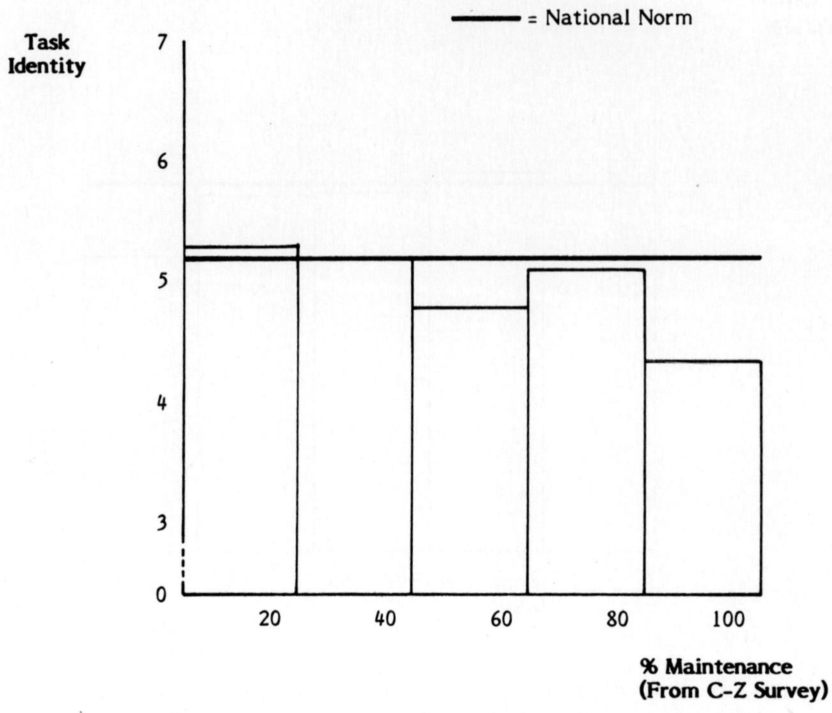

Note: All data was collected as interval data.

Figure 7.2: Task Identity vs. % Maintenance (All Respondents)

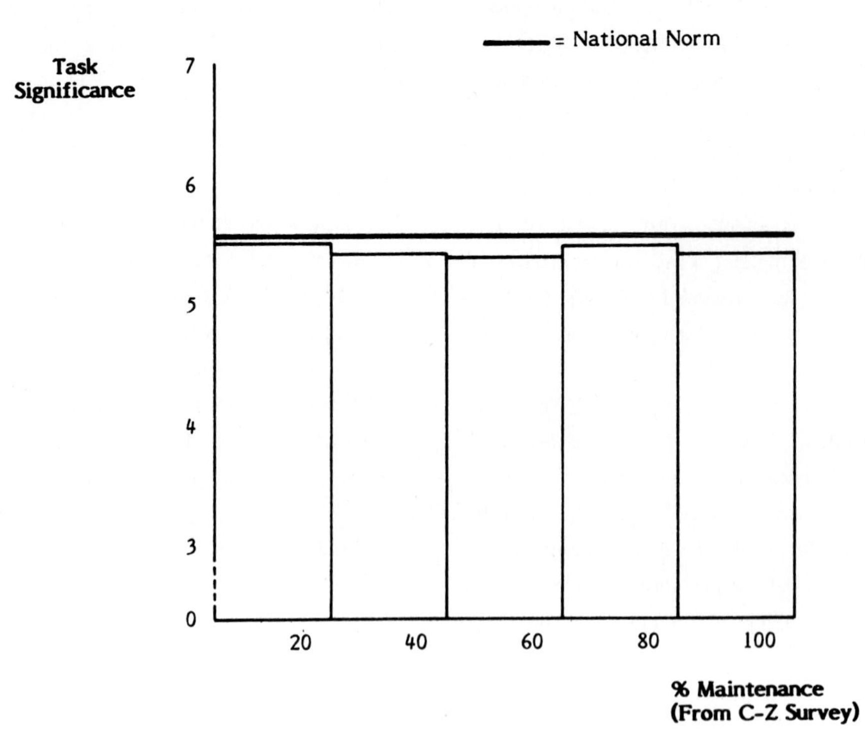

Note: All data was collected as interval data.

Figure 7.3: Task Significance vs. % Maintenance (All Respondents)

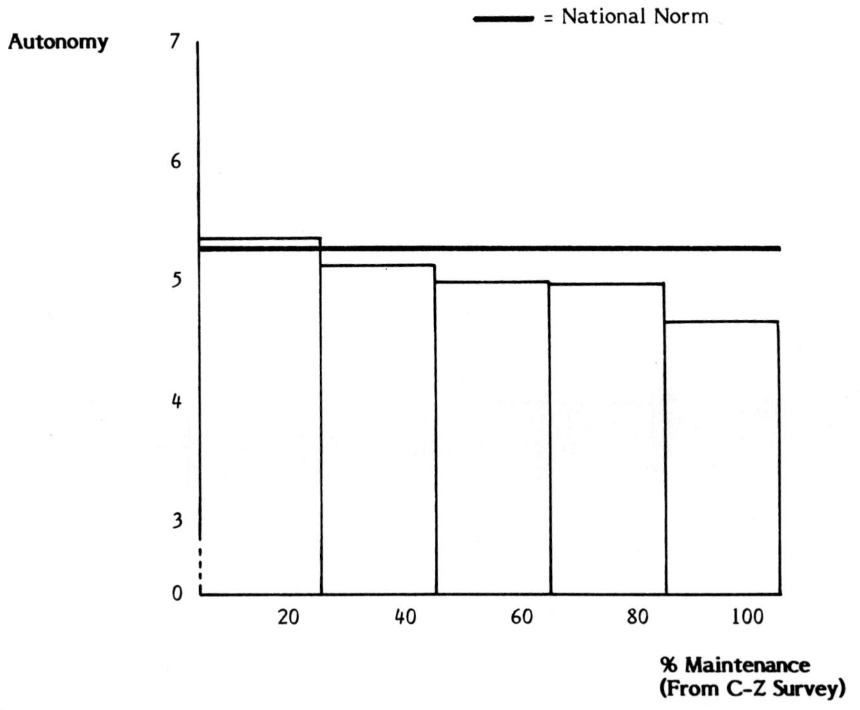

Note: All data was collected as interval data.

Figure 7.4: Autonomy vs. % Maintenance (All Respondents)

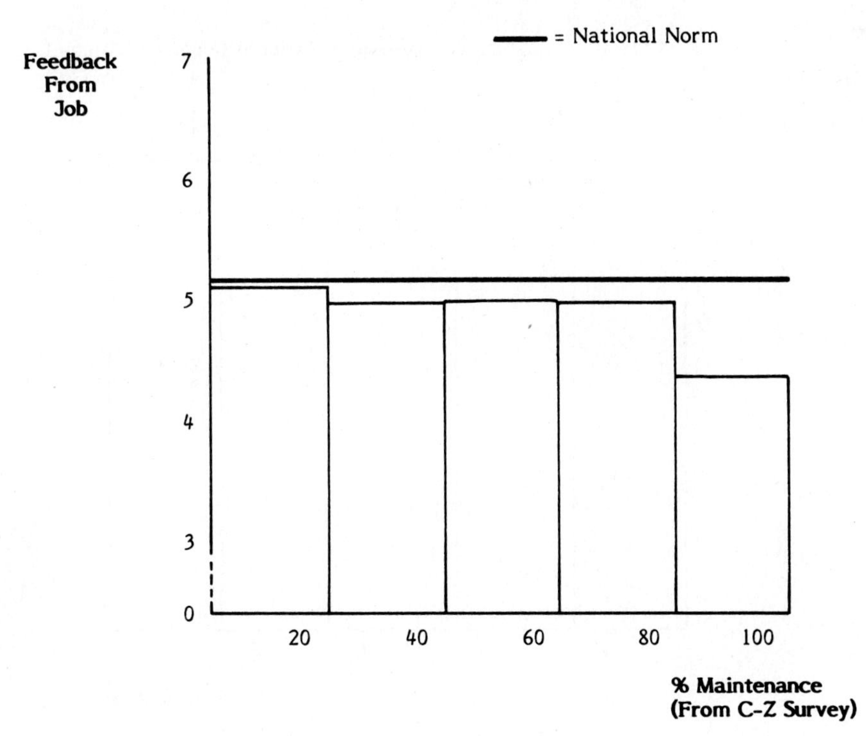

Note: All data was collected as interval data.

Figure 7.5: Feedback From Job vs. % Maintenance (All Respondents)

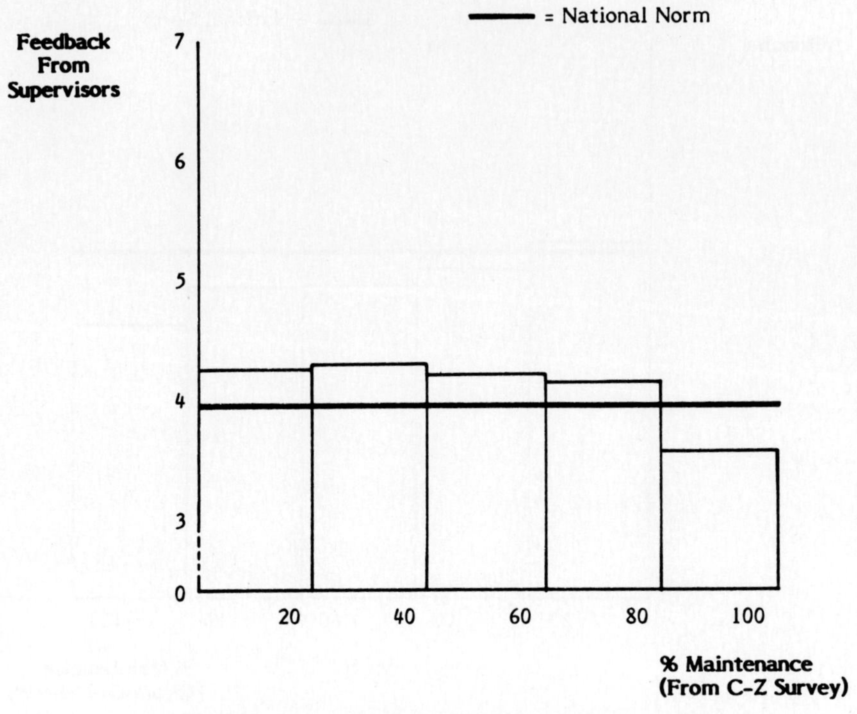

Note: All data was collected as interval data.

Figure 7.6: Feedback From Supervisors vs. % Maintenance (All Respondents)

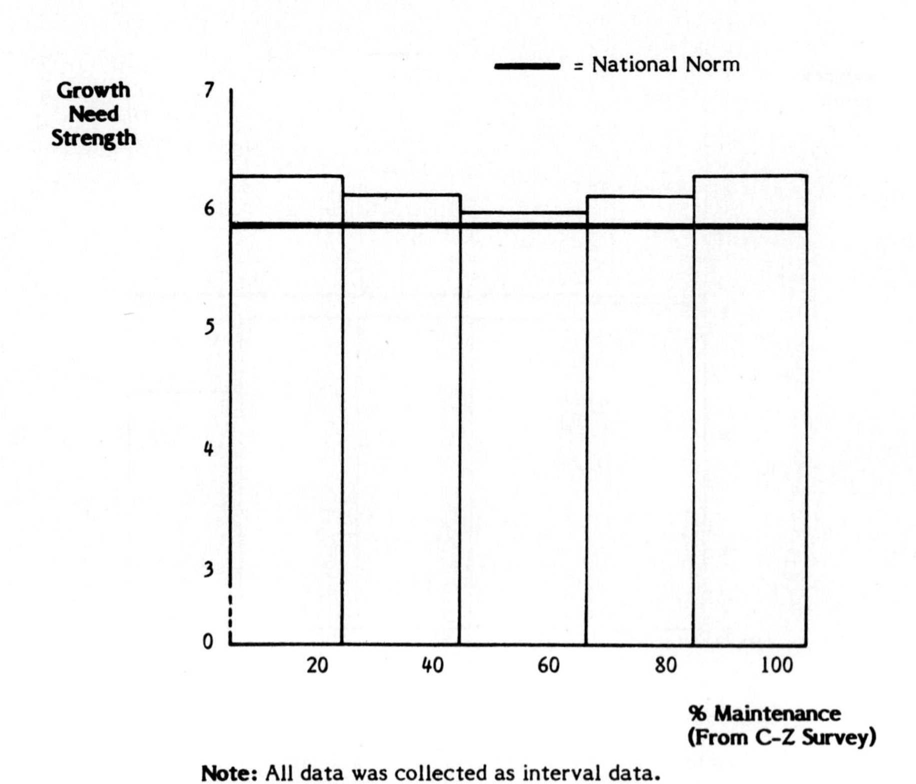

Note: All data was collected as interval data.

Figure 7.7: Growth Need Strength vs. % Maintenance (All Respondents)

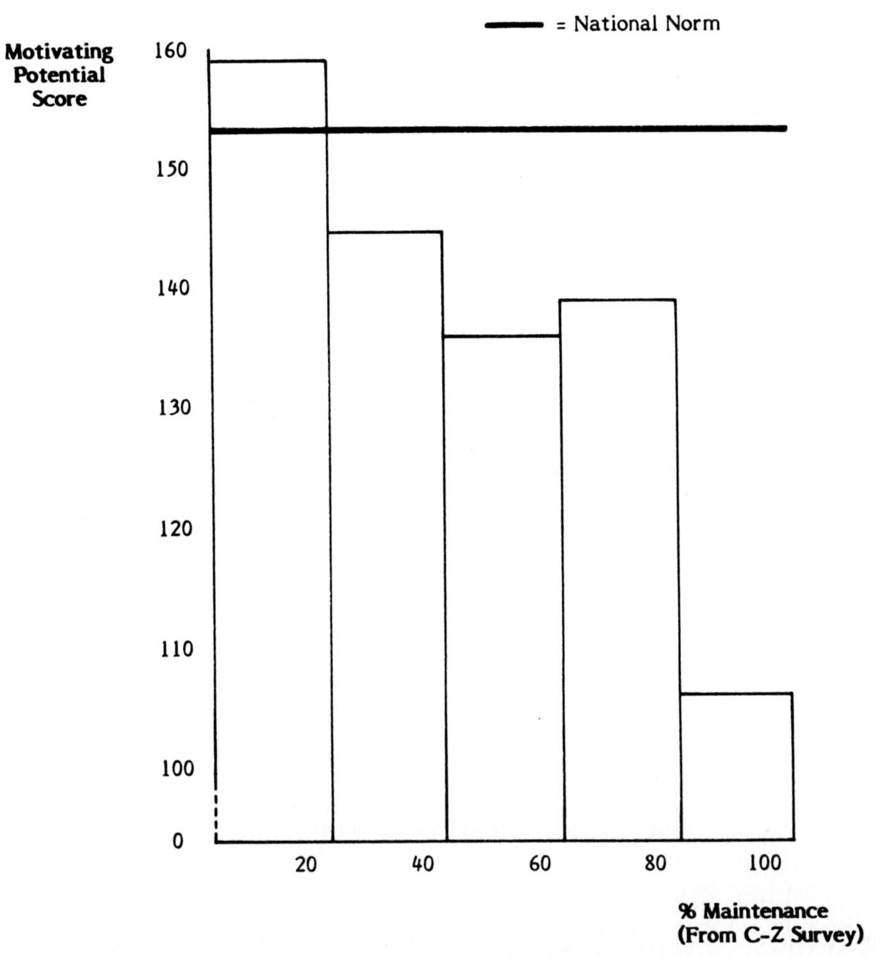

Note: All data was collected as interval data.

Figure 7.8: Motivating Potential Score vs. % Maintenance (All Respondents)

maintenance personnel often work almost in isolation from the rest of the systems group. This isolation even extends to the individual's own manager. Extremely high maintenance efforts are often performed by professionals who hold massive knowledge about the target systems, and they often work relatively independently. As a result, they receive less feedback from their managers. Paradoxically, this same situation results in autonomy being inversely correlated with amount of maintenance (Figure 7.4).

In Figure 7.7, growth need strength is charted against maintenance involvement. Here, a suprising result is that both high and low maintenance individuals exhibit higher than average GNS. This result indicates a possible imbalance between GNS and motivating potential. While this study does not clarify the factors underlying this GNS pattern, it does address the need to match GNS and MPS. The severity of this problem is indicated in Figure 7.8, where MPS is shown to be highly negatively correlated with maintenance activities. This mismatch between GNS and MPS is critical, as noted in several of the cases in Section III.

Finally, in Figure 7.9, the relationship between general satisfaction and maintenance is shown. Again, the negative correlation is present and significant.

These findings are critical, especially considering that the study organizations were specifically chosen for their reputations for well managed and highly productive maintenance efforts. If these organizations exhibit such strong negative correlations between the job variables and maintenance efforts, one would expect that other firms

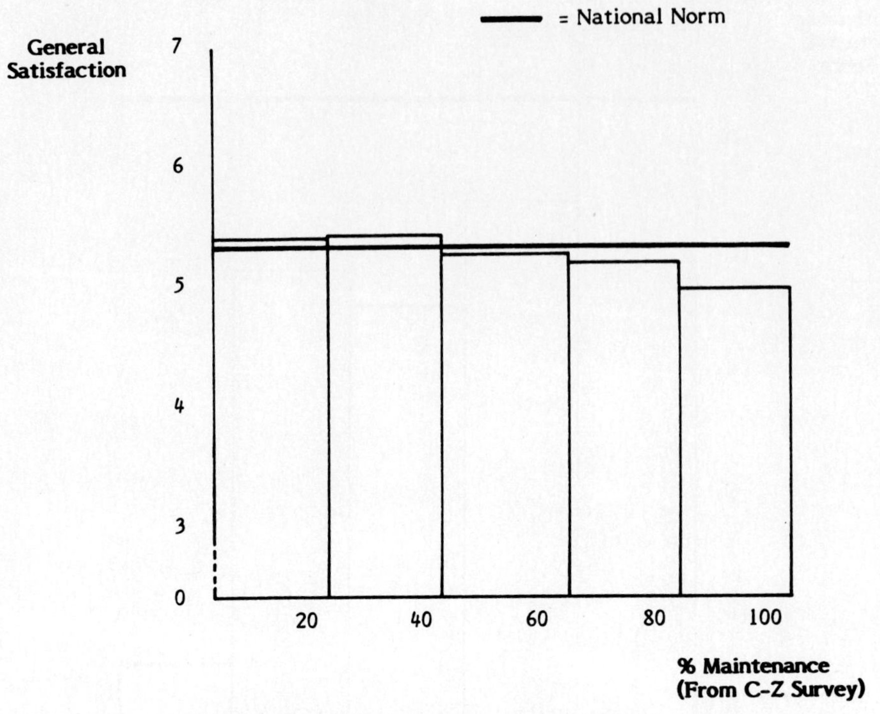

Note: All data was collected as interval data.

Figure 7.9: General Satisfaction vs. % Maintenance (All Respondents)

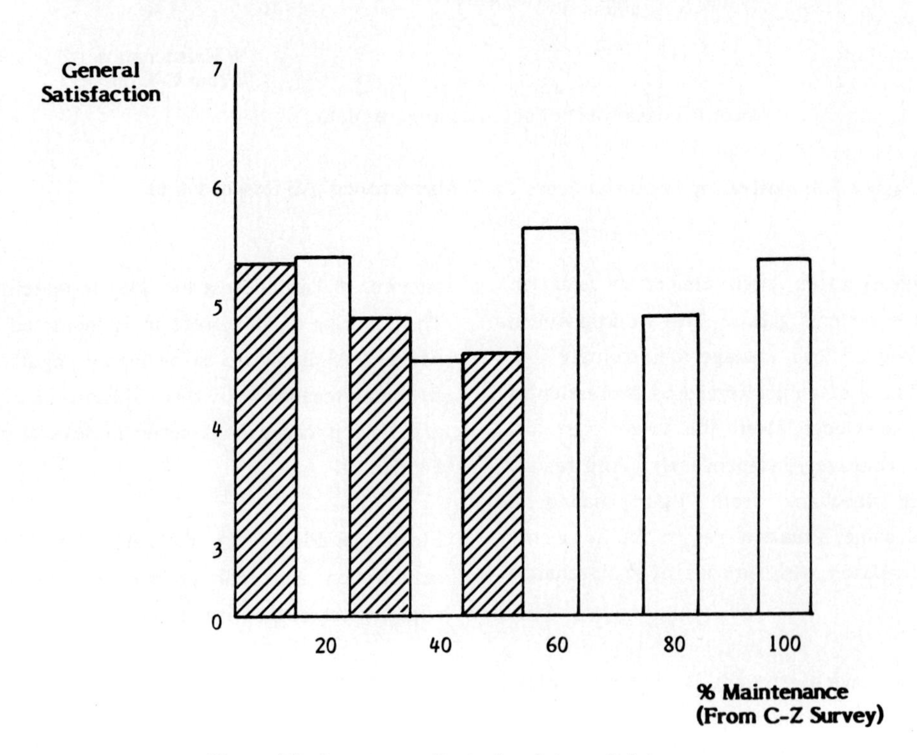

Note: All data was collected as interval data.
-Shaded bar = Fixit, Clear = Enhancement
-No bar = No responses in that catagory.

Figure 7.10: General Satisfaction vs. % Maintenance (Interviewed Subjects)

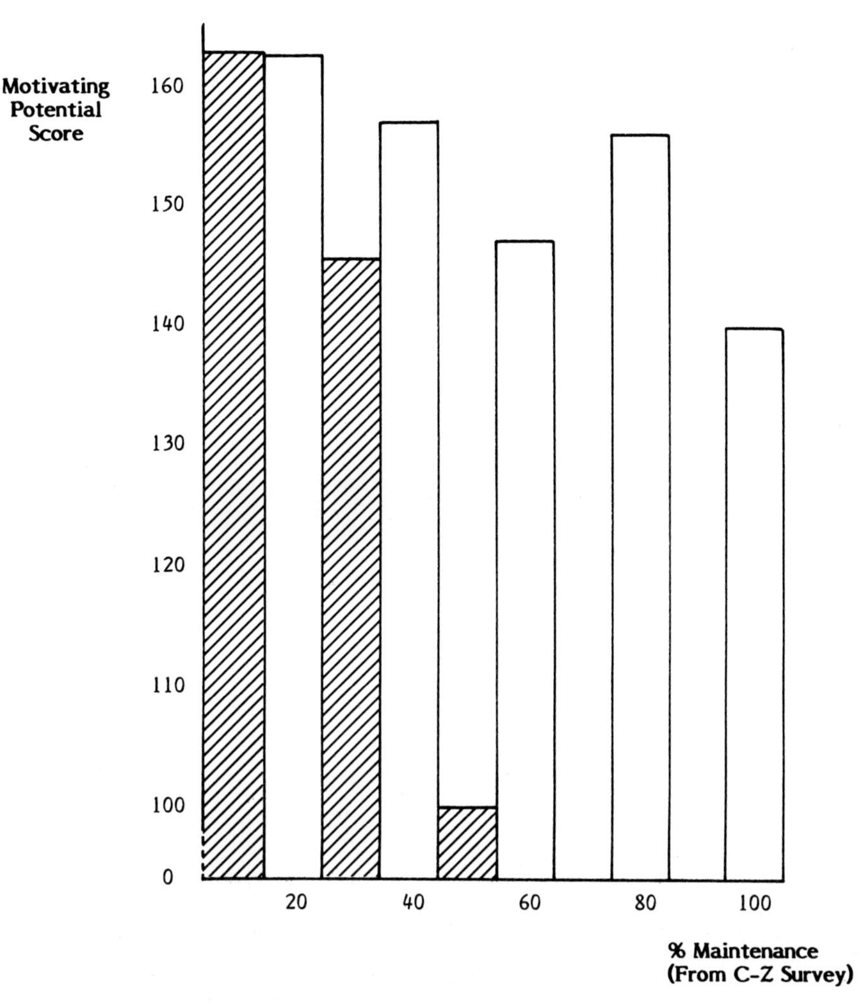

Note: All data was collected as interval data.
-Shaded bar = Fixit, Clear = Enhancement
-No bar = No responses in that catagory.

Figure 7.11: Motivating Potential Score vs. % Maintenance (Interviewed Subjects)

may be experiencing even more negative effects from maintenance assignments. This serves to emphasize the need for careful job design along the lines suggested in Sections II and III.

MAINTENANCE: INTERVIEWED RESPONDENTS

In discussions with both management and non-supervisory personnel, the definition of maintenance was treated at length. No significant difference in the perceptions of the percent of total effort expended on maintenance was found between management and non-supervisory across the entire set of interviewed subjects. Within individual organizations, however, there was wide disagreement.

In order to pursue the relationships between job variables and maintenance activities which were discussed in the last section, a set of linear regression runs was performed with each job variable regressed against a function of both the total amount of maintenance and the amount of fixit activities. In both cases, the amounts were expressed as percentages of employee work activity.

In a set of correlation coefficients which were calculated prior to the regression run, virtually all of the job variables were found to be correlated with the percent of time spent on maintenance in general. All of these correlations were negative and significant at the .01 level. Many were even more highly significant (.001 level). The percent of time spent on enhancement

MAINTENANCE TASK SIZE (LOC)	ALL INTERVIEWED SUBJECTS	EMPLOYEES	MANAGERS
0 to 9	40.298	40.203	40.450
10 to 50	31.250	30.750	32.050
51 to 200	18.433	19.172	17.250
201 and up	10.240	10.234	10.250

FIXIT TASKS

MAINTENANCE TASK SIZE (LOC)	ALL INTERVIEWED SUBJECTS	EMPLOYEES	MANAGERS
0 to 9	15.757	16.844	13.974
10 to 50	22.495	25.813	17.051
51 to 200	28.495	30.313	25.513
201 and up	31.155	26.875	43.462

ENHANCEMENT TASKS

Table 7.1: Estimates of Maintenance Task Size in Lines Of Code (LOC) By Interviewed Subjects

projects significantly correlated negatively with skill variety, task identity, and feedback from supervisors, while the percent of time spent on fixit tasks showed a significant negative correlation with autonomy and feedback from the job.

An interesting observation on the correlations is that enhancement activities are correlated, while not significantly, in a positive direction with task significance and feedback from the job. Fixit tasks, on the other hand, had non-significant positive correlation with skill variety.

In the regression runs, no predictive capabilities were discovered, probably due to the complex multivariate nature of the relationships involved. However, the multiple F statistic values strongly supported the nature and direction of the relationships which were found through simple correlation studies.

These results are summarized graphically in Figures 7.10 and 7.11. When general satisfaction is charted against maintenance after accounting for the breakout of fixit and enhancement activities, it is clear that fixit efforts are far less satisfying than enhancement activities. While no individuals were found with responsibilities which involved fixit activities for more than 60% of their time, the relationships exhibited in the figure indicate that such a job might be particularly low in general satisfaction.

Likewise, Figure 7.11 shows that MPS drops dramatically with the percentage of maintenance activity, while enhancement efforts show less effect on the score.

The remainder of the study involved the analysis of a set of structured questions from the interview sessions. In Table 7.1, the employee and supervisor estimates of maintenance task size in Lines Of Code (LOC) are

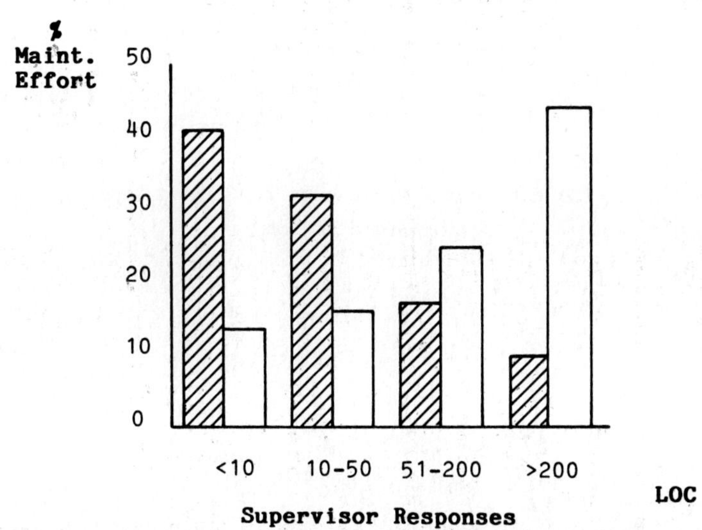

Note: Shaded Bar = Fixit Component
Clear Bar = Enhancement Component

Estimates of Maintenance Task Size in LOC
By Interviewed Subjects
Figure 7.12

MAINTENANCE TASK SIZE (DAYS)	ALL INTERVIEWED SUBJECTS	EMPLOYEES	MANAGERS
Less than ½ day	32.390	28.785	38.250
½ to one day	29.200	29.723	28.350
One to five days	23.133	25.831	18.750
More than five days	14.705	14.892	14.400

FIXIT TASKS

MAINTENANCE TASK SIZE (DAYS)	ALL INTERVIEWED SUBJECTS	EMPLOYEES	MANAGERS
Less than ½ day	9.961	10.359	9.308
½ to one day	17.505	19.484	14.256
One to five days	29.398	31.641	25.718
More than five days	43.233	38.672	50.718

ENHANCEMENT TASKS

Table 7.2: Estimates of Maintenance Task Size in Days By Interviewed Subjects

presented. For fixit tasks, the two groups showed strong agreement on estimated task size. Both managers and employees estimate most fixit efforts to involve the proverbial "few lines of code."

However, managers and employees disagreed sharply on the estimated size of enhancement efforts. While employees perceive enhancement efforts to be evenly distributed with respect to size in LOC, management indicated a belief that most enhancement tasks are quite large. These differences are apparent in Figure 7.12.

In Table 7.2 and Figure 7.13, management and employee estimates of maintenance task size in terms of task duration are presented. Here, there was very little agreement between the two groups. While managers seem to feel that task duration is directly correlated with lines of code, employees see a totally different relationship. Employees see most efforts as taking between one-half to five days, while managers perceive fixit tasks as mostly taking less than one-half day and enhancement tasks mostly taking more than five days.

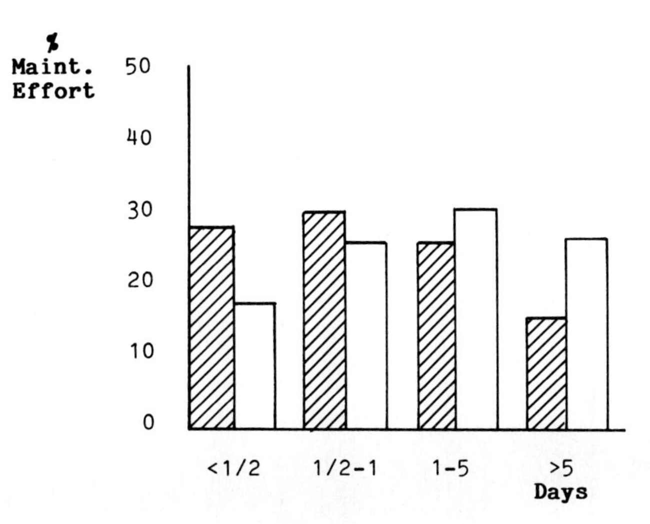

Employee Responses

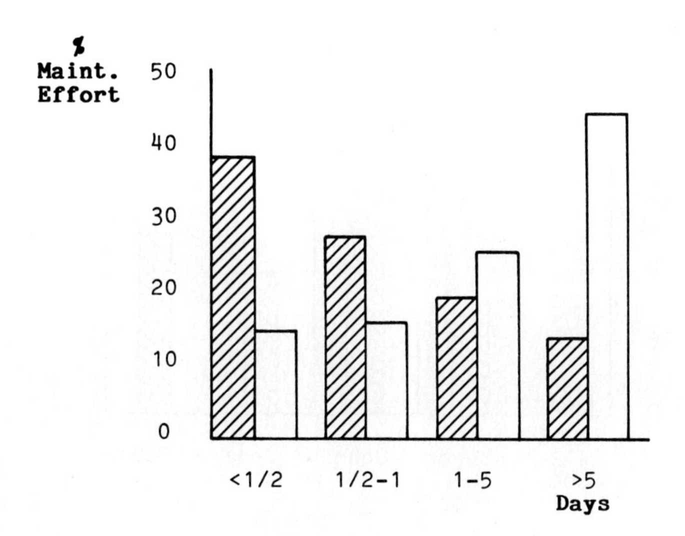

Supervisor Responses

Note: Shaded Bar = Fixit Component
Clear Bar = Enhancement Component

Estimates of Maintenance Task Size in Days
By Interviewed Subjects
Figure 7.13

EVALUATION AREAS	ALL INTERVIEWED SUBJECTS	EMPLOYEES	MANAGERS
All LOC	3.019	2.529	3.850
Error Free LOC	14.148	13.088	15.950
Schedule	48.333	49.706	46.000
Cost	8.639	8.941	8.125
Other	25.722	25.515	26.075

Table 7.3: Perceived Areas of Evaluation of Maintenance Efforts By Interviewed Subjects

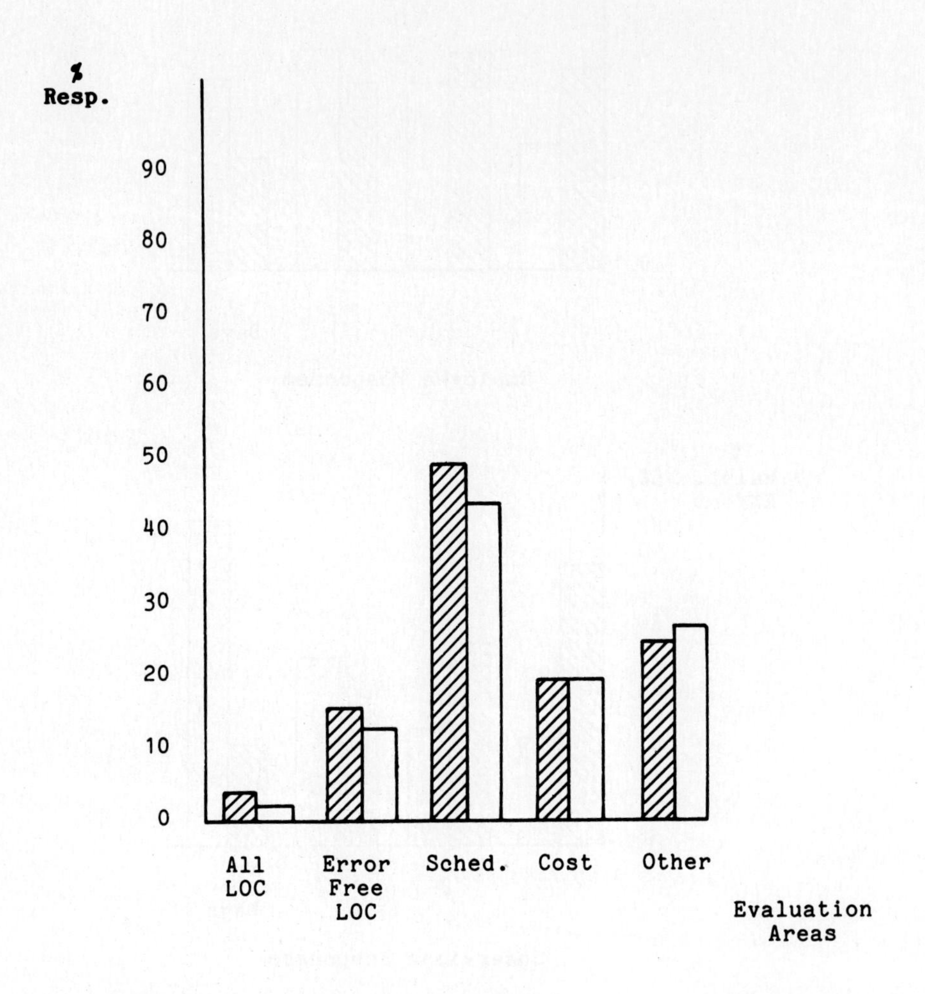

Note: Shaded Bar = Supervisor Responses
Clear Bar = Employee Responses

**Perceived Areas of Evaluation of Maintenance Efforts
By Interviewed Subjects
Figure 7.14**

PROBLEM AREAS	ALL INTERVIEWED SUBJECTS	EMPLOYEES	MANAGERS
Poor Doc.	29.039	29.206	29.200
Poor Design	30.757	33.016	27.200
Poor Code	18.913	19.206	18.450
Time Frame	11.019	8.889	14.375
Other	11.175	9.683	13.525

FIXIT TASKS

PROBLEM AREAS	ALL INTERVIEWED SUBJECTS	EMPLOYEES	MANAGERS
Poor Doc.	25.931	28.387	22.125
Poor Design	30.971	31.516	30.125
Poor Code	17.265	17.194	17.375
Time Frame	14.363	13.548	15.625
Other	11.471	9.355	14.750

ENHANCEMENT TASKS

Table 7.4: Perceived Problem Areas of Maintenance Efforts By Interviewed Subjects

This result is most critical when considered with respect to the next set of results, presented in Table 7.3 and Figure 7.14. Here, both managers and employees agree that the single most important area of evaluation of the maintenance effort involves schedule compliance. If schedule compliance is, indeed, the critical evaluation factor, then agreement on task duration is also critical. This result indicates that the disagreement on task durations may be a factor in some of the difficulties encountered in managing the maintenance effort.

Finally, in Table 7.4 and Figure 7.15, manager and employee responses on perceived problem areas affecting the maintenance effort are presented. Here, the groups were again in agreement. Both management and professionals agree that poor original design and poor documentation account for the majority of problems in this effort. This finding supports statements made in Section III, where Case M discusses the problems of documentation during the maintenance effort. Clearly, these data indicate that organizations should begin to consider maintenance during design and construction and not wait until the maintenance activities actually begin.

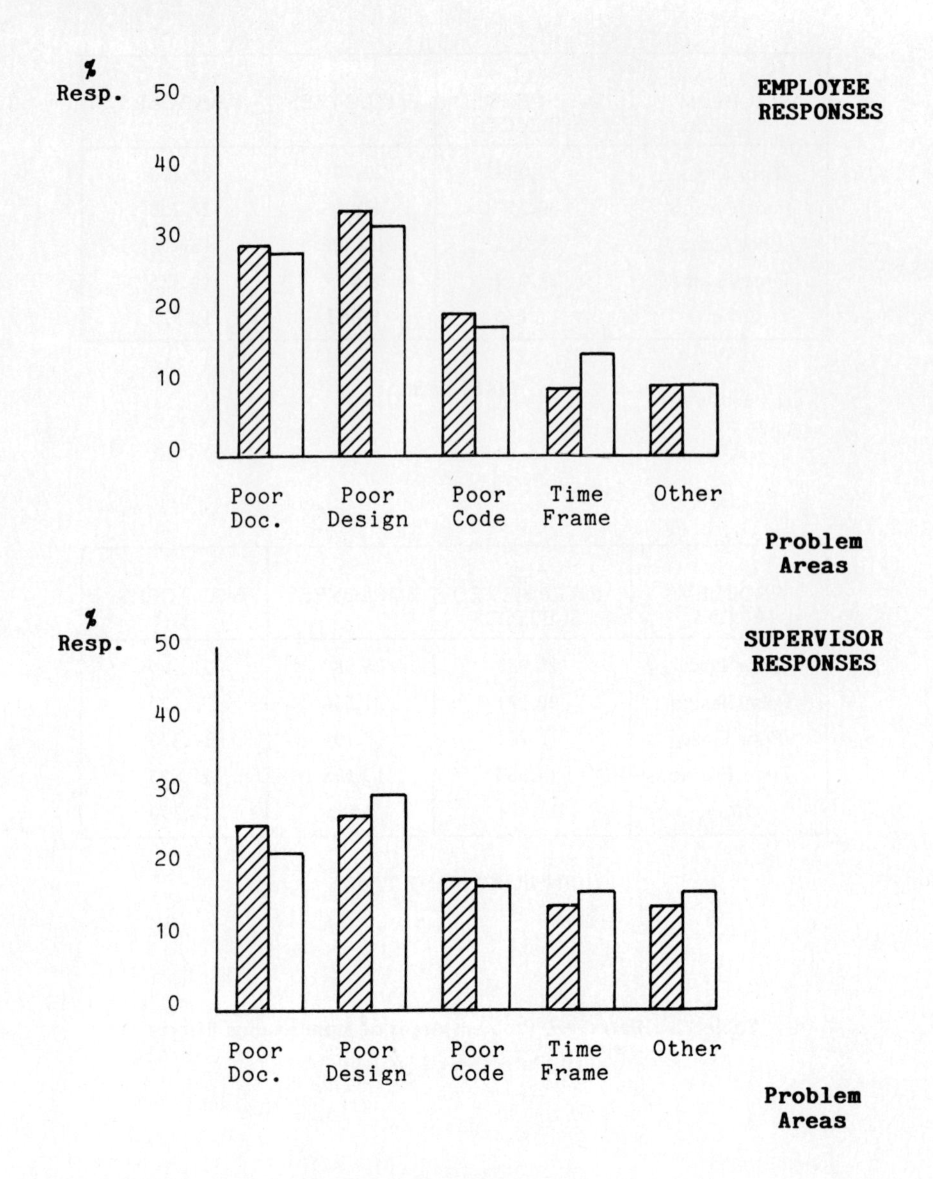

Note: Shaded Bar = Fixit Component
Clear Bar = Enhancement Component

Perceived Problem Areas of Maintenance Efforts By Interviewed Subjects
Figure 7.15

DISCUSSION

The results discussed here shed a great deal of light on the issues involving the maintenance of software systems. First, there is a clear need for the recognition that maintenance involves two primary component efforts. Although prior research has included this issue, it has not been related to motivation and productivity. Our research shows that the two components affect job related measures differently. This implies that job designs which mix the two types of effort will be viewed differently by the employee than a job which contains all of one or the other.

Reference for Chapter 7

1. Swanson, E. "The Dimensions of Maintenance," 2nd International Conference In Software Engineering, Oct. 1976.

APPENDIX I:
DIAGNOSTIC SURVEY INSTRUMENT

BACKGROUND ON THE COUGER-ZAWACKI DIAGNOSTIC SURVEY

The C/Z survey instrument is a diagnostic tool to identify the key factors for motivation and productivity of your entire DP staff.

Professors J. Daniel Couger and Robert A. Zawacki (University of Colorado) proved the reliability and validity of the instrument and published the results in their book Motivating and Managing Computer Personnel (Wiley, 1980). Their data base now contains data on over 6,000 computer personnel. They have developed national norms on motivation for 16 different jobs in the field, including 3 levels of management.

Survey Procedure

The C/Z survey is administered individually to each employee. It takes 25 minutes to complete.

Answer sheets are mailed to C/Z Associates for data entry and computer processing. A report is prepared comparing results in your company to the national norms.

Two types of reporting occur: 1) comparison to national norms by job type, 2) comparison to national norms by organizational unit.

The C/Z survey diagnoses which jobs have low motivating potential and identifies specific ways to improve those jobs. It facilitates career pathing and a skills development program.

The survey instrument measures: 1) job dimensions, 2) satisfaction levels, 3) goal clarity/accomplishment, 4) DP problem areas. Most important, it measures the potential of each job to motivate the specific group of employees holding that job. When a mismatch occurs, clues are provided on how to realign the task assignments. The survey outputs are described below.

1. Key Job Dimensions: Objective characteristics of the job itself.
   - A. Skill Variety: The degree to which a job requires a variety of different activities in carrying out the work, which involve the use of a number of different skills and talents of the employee.
   - B. Task Identity: The degree to which the job requires the completion of a "whole" and identifiable piece of work--i.e., doing a job from beginning to end with a visible outcome.
   - C. Task Significance: The degree to which the job has a substantial impact on the lives or work of other people--whether in the immediate organization or in the external environment.
   - D. Autonomy: The degree to which the job provides substantial freedom, independence, and discretion to the employee in scheduling his/her work and in determining the procedures to be used in carrying it out.
   - E. Feedback from the Job Itself: The degree to which carrying out the work activities required by the job results in the employee obtaining information about the effectiveness of his or her performance.
2. Satisfaction Measures: The private, affective reactions or feelings an employee gets from working on his job.
   - A. General Satisfaction: An overall measure of the degree to which the employee is satisfied and happy in his or her work.
   - B. Internal Work Motivation: The degree to which the employee is self-motivated to perform effectively on the job.
   - C. Specific Satisfactions: These scales tap several specific aspects of the employee's job satisfaction.
     - C1. Pay satisfaction
     - C2. Supervisory satisfaction
     - C3. Satisfaction with co-workers
3. Social Need Strength: This is a measure of the degree to which the employee needs to interact with other employees.
4. Goal Clarity and Accomplishment: These scales measure the degree to which employees understand and accept organizational goals. Further, they tap into the employees' feelings about goal setting participation, goal difficulty, and feedback on goal accomplishment.
5. Problem Areas: This is a measure of the degree to which the following areas are problematic:
   - A. Amount of maintenance being performed.
   - B. Access to the computer.
   - C. Realistic schedules.
6. Experienced Meaningfulness: This scale is a measure

of how worthwhile or important the work is to the employees.

7. Experienced Responsibility: This scale measures the employees' beliefs that they are personally accountable for the outcomes of their efforts.
8. Knowledge of Results: This scale measures the employees' beliefs that they can determine, on some fairly regular basis, whether the outcomes of their work are satisfactory.
9. Individual Growth Need Strength: This scale measures the individual's need for personal accomplishment and for learning and developing beyond his/her present level of knowledge and skills.
10. Motivating Potential Score: A score reflecting the potential of a job for eliciting positive internal work motivation on the part of employees.

BENEFITS FROM USE OF THE COUGER-ZAWACKI DIAGNOSTIC SURVEY

The C-Z Survey pinpoints where improvements can be made in the motivational environment. The following figures show that the typical improvements in any of three areas will pay for the cost of the C-Z Survey.

1. **Productivity Improvements**

   A. Typical Personnel Budget = $2,500,000/year (100 people @ $25,000/year)

   B. 1% improvement in productivity will alone pay for C-Z Survey (1% of $2,500,000 = $25,000)

2. **Reduction in Turnover**

   A. National T/O Average = 25% (34% for application programmers)

   B. Average cost to replace a person = ½ of person's salary

   $2,500,000 x 25% = $625,000/year payroll of persons who leave

   Cost to replace = $312,500 (½ of $625,000)

   Primary costs: Recruiting, training, learning curve, efficiency reduction

   Other costs: Fee to placement agency
   (Standard = 1% per $1K of salary up to 30% of annual salary)
   Mortgage equalization allowance

   C. 5% reduction in turnover will alone pay for the C-Z Survey (5% of $625,000 = $31,250)

   If 2 of your employees decide not to leave, regardless of the total number of employees, that alone will pay for the C-Z Survey
   (2 x $25,000 = $50,000 x .5 replacement cost = $25,000)

3. **Absenteeism Reduction**

   A. Average rate = 8 days per year

   Cost: $\frac{\$25{,}000 \text{ average salary}}{245 \text{ work days/year}} = \$102$ per day

   B. Motivated employees can be expected to be absent less often.
   Reduction from 8 to 5 days average alone pays for the C-Z Survey
   ($102/day x 3 days/year reduction x 100 people = $30,600/year)

* *

CONCLUSION:

If increased motivation leads to a productivity increase of only 1 percent or if only 2 employees decide not to turnover—the survey will pay for itself.

* *

For Additional Information:

Cost of the survey may be obtained by writing C-Z Survey, Box 7345, Colorado Springs, CO 80933

APPENDIX II:
STRUCTURED INTERVIEW QUESTIONNAIRE ON MOTIVATION

STRUCTURED INTERVIEW QUESTIONNAIRE ON MOTIVATION

_______ Code Initials   _______ % of Maintenance

PART I

Qualitative Questions

Five (5) Core Job Dimensions

The respondents from this organization have scored as follows on the 5 Core Job Dimensions:

	High	Average	Low
Skill Variety			
Task Identity			
Task Significance			
Autonomy			
Feedback from the Job			

To the Interviewee:

In examining the data from the organization and comparing it with national survey data, we noticed that your organization has scored well (above/below) the national norm in the dimension(s) of ______________________ (Here the researcher may need to ______________________ define the dimension briefly.)*

Can you tell me what you believe might have caused or contributed to these findings?

Dimension ____

____

____

____

____

____

____

____

____

____

____

Do you have any thoughts as to why we found ______________________ to be (high/low)?

Dimension ____

____

____

____

____

____

____

____

____

____

____

Do you have any thoughts as to why we found ______________________ to be (high/low)?

Dimension ____

____

____

____

____

____

____

____

____

____

____

Do you have any thoughts as to why we found ______________________ to be (high/low)?

Dimension ____

____

____

____

____

____

____

____

____

____

____

PART II

OK, now I'd like to ask you to respond to ten (10) questions in a very short questionnaire that also taps your perceptions about your job. I'd then like to ask you some questions about your responses. Again, these answers are to be kept in a completely anonymous fashion. Our use of the code initial simply allows us to keep the several parts of our data together for later analysis. No one in management will see your response. Data will be pooled and presented to management only in summary form.

Here is the questionnaire. Please put your code initials (the initials of your mother's maiden name) at the top.

_______ Code Initials

JOB PERCEPTION QUESTIONNAIRE

Please respond to each of the ten (10) questions by √ or X below as you usually see things regarding your job.

	Very False	False	Somewhat False	Neither True or False	Somewhat True	True	Very True
1. I know what my responsibilities are.	____	____	____	____	____	____	____
2. The company gives me recognition for producing high quality output.	____	____	____	____	____	____	____
3. I have to do things that should be done differently.	____	____	____	____	____	____	____
4. I know exactly what is expected of me.	____	____	____	____	____	____	____
5. The company gives me recognition for getting my job done on time.	____	____	____	____	____	____	____
6. I have to buck a rule or policy in order to carry out an assignment.	____	____	____	____	____	____	____

	Very False	False	Somewhat False	Neither True or False	Somewhat True	True	Very True
7. I know that I have divided my time properly.	____	____	____	____	____	____	____
8. The company gives no recognition for producing a high quality of output.	____	____	____	____	____	____	____
9. I receive incompatible requests from two or more people.	____	____	____	____	____	____	____
10. Getting the job done on time leads to job security here.	____	____	____	____	____	____	____

To Researcher:

Quickly score the three (3) dimensions with the template.

Dimension	Score (Circle as Appropriate) Low	High
I Role Clarity	Below 9 Role not clear	Over 15 Role very clear
II Clarity of Rewards Linkages	Below 12 Not Clear	Over 20 Very Clear
III Role Conflict	Below 9 Little role Conflict	Over 15 Much conflict in roles

Assess above, then select from I, II, III below.

To Interviewee:

I. It looks as though you have some pretty strong feelings about the role or roles you are asked to occupy. Your answers tend to indicate that you have a very clear (unclear) set of perceptions about what is expected of you in your job. Can you elaborate a bit?

____

____

____

____

____

____

____

II. Your responses tend to indicate that the linkage between performance and rewards—as you see them, at least—are pretty clear (unclear). Could you discuss why this might be?

____

____

____

____

____

____

____

III. In looking at your answers on the questionnaire, it also looks like you find much (very little) conflict in the demands made of you on this job.
Can you describe why this might be?

____

____

____

____

____

____

____

PART III

For the following questions, you are asked to weight each alternative, using percentages that total to 100%. Round your figures to increments of 5%.

1. What percent of your time is spent in each of the following types of jobs?

	% of Total
Jobs requiring less than 10 lines of code	______
Jobs requiring 10 to 50 lines of code	______
Jobs requiring 51 to 200 lines of code	______
Jobs requiring over 200 lines of code	______
TOTAL	100%

2. Identify the length of time you spend on jobs.

	% of Total
Jobs requiring less than 4 hours	______
Jobs requiring 4 to 8 hours	______
Jobs requiring 2 to 5 days	______
Jobs requiring over 5 days	______
TOTAL	100%

3. Weight the importance of the following measures used by your supervisor in evaluating your productivity:

	% of Total
Lines of code produced	______
Lines of error-free code produced	______
Schedule compliance	______
Reduction in computer processing cost	______
Other: ______________	______
TOTAL	100%

4. Weight the following factors for determining who works primarily on maintenance tasks:

	% of Total
Employee preference for maintenance work	________
Department policy (everyone rotates from new system development to maintenance)	________
Training approach for new employees	________
Availability (employee must maintain a system until a new development project is available)	________
As a penalty (maintenance assigned to whom the boss is "down on")	________
Other: ____________________	________
TOTAL	100%

5. Weight the degree of severity of the following problems in maintaining systems:

	% of Total
Poor documentation	
Poor design of original program	________
Inefficient coding of original program	________
Unrealistic schedules for maintenance	________
Other: ____________________	________
TOTAL	100%

6. To what extent are promotion and merit raises based on:

	% of Total
Quantitative measures of productivity	________
Qualitative judgments	________
By supervisor	________
By others	________
Other bases: ____________________	________
TOTAL	100%

7. Are any special considerations made for personnel whose jobs are over 50% maintenance?

1) Flextime ____________

2) Pay differential (e.g. overtime) ____________

3) Less pressure on schedule ____________

4) More training time ____________

5) Faster promotion ____________

6) Special equipment (your own terminal) ____________

7) Other: ______________________ ____________

Concluding Open Ended Question:

____

____

____

____

____

____

____

____

____

____

____

____

____

____

Note: Respondent should now review entire form and return it to you.

APPENDIX III: RESEARCH CASES

The following set of figures contains the actual analysis results from the eight organizations which were involved in the maintenance study. These figures are not developed in detail through discussions, because the most important results have already been incorporated into the application cases in Chapter 5. The actual data from the research effort is presented in order to support the main body of the report.

The organization of the research cases is as follows. Each organization, referenced by the number assigned to it during the study, is presented in two pages of tables. The first page details the demographics of the organization, comparing its scores on the demographic variables with <u>both</u> the national norms and the set of data from all respondents of all study firms. The second page of each research case compares the job variable scores for the firm to the national norms and to the set of all respondents.

VARIABLE	NATIONAL DATA		FIRM 42 DATA		STAT TEST RESULTS
	MEAN (M1)	STD.DEV. (V1)	MEAN (M2)	STD.DEV. (V2)	
SEX	1.262	.450	1.375	.495	
AGE	2.939	.913	2.833	1.049	
EDUCATION	4.594	.702	4.750	.989	V2 GT V1 (.01)
MARITAL STATUS	1.626	.487	1.625	.576	
YEARS WITH FIRM	2.728	1.419	1.708	.751	M2 LT M1 (.01) V2 GT V1 (.01)

VARIABLE	ALL RESPOND.		FIRM 42 DATA		STAT TEST RESULTS
	MEAN (M1)	STD.DEV. (V1)	MEAN (M2)	STD.DEV. (V2)	
SEX	1.342	.479	1.375	.495	
AGE	2.813	.945	2.833	1.049	
EDUCATION	4.632	1.134	4.750	.989	
MARITAL STATUS	1.591	.544	1.625	.576	
YEARS WITH FIRM	2.804	1.695	1.708	.751	M2 LT M1 (.01) V2 LT V1 (.01)

For National Data, N = 709
For All Respondents, N = 555
For This Firm, N = 24

NOTE...Significance levels are in parentheses in Results Column.

Variables are scored as follows:

SEX : Male = 1; Female = 2.
AGE : Under 20 = 1; Twenties = 2; Thirties = 3; Forties = 4; Fifties = 5; Sixties = 6.
EDUC: Grade = 1; Some High = 2; High = 3; Some College = 4; B.A. or B.S. = 5; M.A. or M.S. = 6; Ph.D. = 7.
MSTAT: Single = 1; Married = 2.
YEARS: 1 or less = 1; 1 to 4 = 2; 4 to 8 = 3; 8 to 12 = 4; 12 to 16 = 5; Over 16 = 6.

Table III.1: Firm 42 Demographics

VARIABLE	NATIONAL DATA		FIRM 42 DATA		STAT TEST RESULTS
	MEAN (M1)	STD.DEV. (V1)	MEAN (M2)	STD.DEV. (V2)	
SKILL VARIETY	5.408	1.106	4.736	1.150	M2 LT M1 (.01)
TASK IDENTITY	5.205	1.110	4.236	1.546	M2 LT M1 (.01) V2 GT V1 (.01)
TASK SIGNIFICANCE	5.605	1.176	4.972	1.624	M2 LT M1 (.05) V2 GT V1 (.01)
AUTONOMY	5.290	1.098	5.153	1.355	V2 GT V1 (.05)
FEEDBACK FROM JOB	5.133	1.066	4.656	.975	M2 LT M1 (.05)
FDBK FROM SUPERVISORS	3.966	1.559	4.486	1.532	
GROWTH NEED STRENGTH	5.905	.994	6.361	.522	M2 GT M1 (.01) V2 LT V1 (.01)
MPS	153.581	62.464	118.792	57.474	M2 LT M1 (.01)

VARIABLE	ALL RESPONDENTS		FIRM 42 DATA		STAT TEST RESULTS
	MEAN (M1)	STD.DEV. (V1)	MEAN (M2)	STD.DEV. (V2)	
SKILL VARIETY	5.525	.983	4.736	1.150	M2 LT M1 (.01)
TASK IDENTITY	5.102	1.150	4.236	1.546	M2 LT M1 (.01) V2 GT V1 (.01)
TASK SIGNIFICANCE	5.486	1.146	4.972	1.624	V2 GT V1 (.01)
AUTONOMY	5.189	1.072	5.153	1.355	V2 GT V1 (.05)
FEEDBACK FROM JOB	5.005	1.104	4.656	.975	M2 LT M1 (.05)
FDBK FROM SUPERVISORS	4.184	1.549	4.486	1.532	
GROWTH NEED STRENGTH	6.217	.771	6.361	.522	V2 LT V1 (.01)
MPS	146.609	64.292	118.792	57.474	M2 LT M1 (.05)

All variables except MPS are scored on a 1 to 7 scale.

Table III.2: Firm 42 Job Variables

VARIABLE	NATIONAL DATA		FIRM 43 DATA		STAT TEST RESULTS
	MEAN (M1)	STD.DEV. (V1)	MEAN (M2)	STD.DEV. (V2)	
SEX	1.262	.450	1.390	.494	
AGE	2.939	.913	3.098	.800	
EDUCATION	4.594	.702	4.366	.698	M2 LT M1 (.05)
MARITAL STATUS	1.626	.487	1.585	.591	V2 GT V1 (.05)
YEARS WITH FIRM	2.728	1.419	2.927	1.456	

VARIABLE	ALL RESPOND.		FIRM 43 DATA		STAT TEST RESULTS
	MEAN (M1)	STD.DEV. (V1)	MEAN (M2)	STD.DEV. (V2)	
SEX	1.342	.479	1.390	.494	
AGE	2.813	.945	3.098	.800	M2 GT M1 (.05)
EDUCATION	4.632	1.134	4.366	.698	M2 LT M1 (.05) V2 LT V1 (.01)
MARITAL STATUS	1.591	.544	1.585	.591	
YEARS WITH FIRM	2.804	1.695	2.927	1.456	

For National Data, N = 709
For All Respondents, N = 555
For This Firm, N = 41

NOTE...Significance levels are in parentheses in Results Column.

Variables are scored as follows:

SEX : Male = 1; Female = 2.
AGE : Under 20 = 1; Twenties = 2; Thirties = 3; Forties = 4; Fifties = 5; Sixties = 6.
EDUC: Grade = 1; Some High = 2; High = 3; Some College = 4; B.A. or B.S. = 5; M.A. or M.S. = 6; Ph.D. = 7.
MSTAT: Single = 1; Married = 2.
YEARS: 1 or less = 1; 1 to 4 = 2; 4 to 8 = 3; 8 to 12 = 4; 12 to 16 = 5; Over 16 = 6.

Table III.3: Firm 43 Demographics

VARIABLE	NATIONAL DATA		FIRM 43 DATA		STAT TEST RESULTS
	MEAN (M1)	STD.DEV. (V1)	MEAN (M2)	STD.DEV. (V2)	
SKILL VARIETY	5.408	1.106	5.593	1.037	
TASK IDENTITY	5.205	1.110	5.520	.969	M2 GT M1 (.05)
TASK SIGNIFICANCE	5.605	1.176	5.715	1.092	
AUTONOMY	5.290	1.098	5.407	1.076	
FEEDBACK FROM JOB	5.133	1.066	4.890	1.044	
FDBK FROM SUPERVISORS	3.966	1.559	4.220	1.629	
GROWTH NEED STRENGTH	5.905	.994	6.297	.755	M2 GT M1 (.01) V2 LT V1 (.05)
MPS	153.581	62.464	154.314	66.031	

VARIABLE	ALL RESPONDENTS		FIRM 43 DATA		STAT TEST RESULTS
	MEAN (M1)	STD.DEV. (V1)	MEAN (M2)	STD.DEV. (V2)	
SKILL VARIETY	5.525	.983	5.593	1.037	
TASK IDENTITY	5.102	1.150	5.520	.969	M2 GT M1 (.01)
TASK SIGNIFICANCE	5.486	1.146	5.715	1.092	
AUTONOMY	5.189	1.072	5.407	1.076	
FEEDBACK FROM JOB	5.005	1.104	4.890	1.044	
FDBK FROM SUPERVISORS	4.184	1.549	4.220	1.629	
GROWTH NEED STRENGTH	6.217	.771	6.297	.755	
MPS	146.609	64.292	154.314	66.031	

All variables except MPS are scored on a 1 to 7 scale.

Table III.4: Firm 43 Job Variables

VARIABLE	NATIONAL DATA		FIRM 45 DATA		STAT TEST RESULTS
	MEAN (M1)	STD.DEV. (V1)	MEAN (M2)	STD.DEV. (V2)	
SEX	1.262	.450	1.505	.503	M2 GT M1 (.01) V2 GT V1 (.05)
AGE	2.939	.913	2.451	.637	M2 LT M1 (.01) V2 LT V1 (.01)
EDUCATION	4.594	.702	4.714	.886	V2 GT V1 (.01)
MARITAL STATUS	1.626	.487	1.549	.543	V2 GT V1 (.05)
YEARS WITH FIRM	2.728	1.419	2.253	1.387	M2 LT M1 (.01)

VARIABLE	ALL RESPOND.		FIRM 45 DATA		STAT TEST RESULTS
	MEAN (M1)	STD.DEV. (V1)	MEAN (M2)	STD.DEV. (V2)	
SEX	1.342	.479	1.505	.503	M2 GT M1 (.01)
AGE	2.813	.945	2.451	.637	M2 LT M1 (.01) V2 LT V1 (.01)
EDUCATION	4.632	1.134	4.714	.886	V2 LT V1 (.01)
MARITAL STATUS	1.591	.544	1.549	.543	
YEARS WITH FIRM	2.804	1.695	2.253	1.387	M2 LT M1 (.01) V2 LT V1 (.01)

For National Data, N = 709
For All Respondents, N = 555
For This Firm, N = 91

NOTE...Significance levels are in parentheses in Results Column.

Variables are scored as follows:

SEX : Male = 1; Female = 2.
AGE : Under 20 = 1; Twenties = 2; Thirties = 3; Forties = 4; Fifties = 5; Sixties = 6.
EDUC: Grade = 1; Some High = 2; High = 3; Some College = 4; B.A. or B.S. = 5; M.A. or M.S. = 6; Ph.D. = 7.
MSTAT: Single = 1; Married = 2.
YEARS: 1 or less = 1; 1 to 4 = 2; 4 to 8 = 3; 8 to 12 = 4; 12 to 16 = 5; Over 16 = 6.

Table III.5: Firm 45 Demographics

VARIABLE	NATIONAL DATA		FIRM 45 DATA		STAT TEST RESULTS
	MEAN (M1)	STD.DEV. (V1)	MEAN (M2)	STD.DEV. (V2)	
SKILL VARIETY	5.408	1.106	5.359	.923	V2 LT V1 (.01)
TASK IDENTITY	5.205	1.110	4.623	1.189	M2 LT M1 (.01)
TASK SIGNIFICANCE	5.605	1.176	5.359	1.194	M2 LT M1 (.05)
AUTONOMY	5.290	1.098	4.751	1.130	M2 LT M1 (.01)
FEEDBACK FROM JOB	5.133	1.066	4.640	1.273	M2 LT M1 (.01) V2 GT V1 (.01)
FDBK FROM SUPERVISORS	3.966	1.559	3.821	1.562	
GROWTH NEED STRENGTH	5.905	.994	6.163	.706	M2 GT M1 (.01) V2 LT V1 (.01)
MPS	153.581	62.464	120.626	62.005	M2 LT M1 (.01)

VARIABLE	ALL RESPONDENTS		FIRM 45 DATA		STAT TEST RESULTS
	MEAN (M1)	STD.DEV. (V1)	MEAN (M2)	STD.DEV. (V2)	
SKILL VARIETY	5.525	.983	5.359	.923	
TASK IDENTITY	5.102	1.150	4.623	1.189	M2 LT M1 (.01)
TASK SIGNIFICANCE	5.486	1.146	5.359	1.194	
AUTONOMY	5.189	1.072	4.751	1.130	M2 LT M1 (.01)
FEEDBACK FROM JOB	5.005	1.104	4.640	1.273	M2 LT M1 (.01) V2 GT V1 (.01)
FDBK FROM SUPERVISORS	4.184	1.549	3.821	1.562	M2 LT M1 (.05)
GROWTH NEED STRENGTH	6.217	.771	6.163	.706	
MPS	146.609	64.292	120.626	62.005	M2 LT M1 (.01)

All variables except MPS are scored on a 1 to 7 scale.

Table III.6: Firm 45 Job Variables

VARIABLE	NATIONAL DATA		FIRM 46 DATA		STAT TEST RESULTS
	MEAN (M1)	STD.DEV. (V1)	MEAN (M2)	STD.DEV. (V2)	
SEX	1.262	.450	1.611	.502	M2 GT M1 (.01)
AGE	2.939	.913	2.278	.752	M2 LT M1 (.01)
EDUCATION	4.594	.702	5.278	.669	M2 GT M1 (.01)
MARITAL STATUS	1.626	.487	1.611	.502	
YEARS WITH FIRM	2.728	1.419	2.000	1.029	M2 LT M1 (.01)

VARIABLE	ALL RESPOND.		FIRM 46 DATA		STAT TEST RESULTS
	MEAN (M1)	STD.DEV. (V1)	MEAN (M2)	STD.DEV. (V2)	
SEX	1.342	.479	1.611	.502	M2 GT M1 (.01)
AGE	2.813	.945	2.278	.752	M2 LT M1 (.01)
EDUCATION	4.632	1.134	5.278	.669	M2 GT M1 (.01) V2 LT V1 (.01)
MARITAL STATUS	1.591	.544	1.611	.502	
YEARS WITH FIRM	2.804	1.695	2.000	1.029	M2 LT M1 (.01) V2 LT V1 (.01)

For National Data, N = 709
For All Respondents, N = 555
For This Firm, N = 18

NOTE...Significance levels are in parentheses in Results Column.

Variables are scored as follows:
SEX : Male = 1; Female = 2.
AGE : Under 20 = 1; Twenties = 2; Thirties = 3; Forties = 4; Fifties = 5; Sixties = 6.
EDUC: Grade = 1; Some High = 2; High = 3; Some College = 4; B.A. or B.S. = 5; M.A. or M.S. = 6; Ph.D. = 7.
MSTAT: Single = 1; Married = 2.
YEARS: 1 or less = 1; 1 to 4 = 2; 4 to 8 = 3; 8 to 12 = 4; 12 to 16 = 5; Over 16 = 6.

Table III.7: Firm 46 Demographics

VARIABLE	NATIONAL DATA		FIRM 46 DATA		STAT TEST RESULTS
	MEAN (M1)	STD.DEV. (V1)	MEAN (M2)	STD.DEV. (V2)	
SKILL VARIETY	5.408	1.106	5.759	.995	
TASK IDENTITY	5.205	1.110	5.167	.958	
TASK SIGNIFICANCE	5.605	1.176	5.722	.945	
AUTONOMY	5.290	1.098	5.426	1.107	
FEEDBACK FROM JOB	5.133	1.066	5.653	.936	M2 GT M1 (.05)
FDBK FROM SUPERVISORS	3.966	1.559	4.648	1.639	M2 GT M1 (.05)
GROWTH NEED STRENGTH	5.905	.994	6.185	.618	M2 GT M1 (.05) V2 LT V1 (.05)
MPS	153.581	62.464	173.670	52.859	

VARIABLE	ALL RESPONDENTS		FIRM 46 DATA		STAT TEST RESULTS
	MEAN (M1)	STD.DEV. (V1)	MEAN (M2)	STD.DEV. (V2)	
SKILL VARIETY	5.525	.983	5.759	.995	
TASK IDENTITY	5.102	1.150	5.167	.958	
TASK SIGNIFICANCE	5.486	1.146	5.722	.945	
AUTONOMY	5.189	1.072	5.426	1.107	
FEEDBACK FROM JOB	5.005	1.104	5.653	.936	M2 GT M1 (.01)
FDBK FROM SUPERVISORS	4.184	1.549	4.648	1.639	
GROWTH NEED STRENGTH	6.217	.771	6.185	.618	
MPS	146.609	64.292	173.670	52.859	M2 GT M1 (.05)

All variables except MPS are scored on a 1 to 7 scale.

Table III.8: Firm 46 Job Variables

VARIABLE	NATIONAL DATA		FIRM 48 DATA		STAT TEST RESULTS
	MEAN (M1)	STD.DEV. (V1)	MEAN (M2)	STD.DEV. (V2)	
SEX	1.262	.450	1.095	.436	M2 LT M1 (.05)
AGE	2.939	.913	2.762	1.044	
EDUCATION	4.594	.702	4.286	1.384	V2 GT V1 (.01)
MARITAL STATUS	1.626	.487	1.429	.676	V2 GT V1 (.01)
YEARS WITH FIRM	2.728	1.419	1.857	.854	M2 LT M1 (.01) V2 LT V1 (.01)

VARIABLE	ALL RESPOND.		FIRM 48 DATA		STAT TEST RESULTS
	MEAN (M1)	STD.DEV. (V1)	MEAN (M2)	STD.DEV. (V2)	
SEX	1.342	.479	1.095	.436	M1 LT M2 (.01)
AGE	2.813	.945	2.762	1.044	
EDUCATION	4.632	1.134	4.286	1.384	
MARITAL STATUS	1.591	.544	1.429	.676	
YEARS WITH FIRM	2.804	1.695	1.857	.854	M2 LT M1 (.01) V2 LT V1 (.01)

For National Data, N = 709
For All Respondents, N = 555
For This Firm, N = 21

NOTE...Significance levels are in parentheses in Results Column.

Variables are scored as follows:

- SEX : Male = 1; Female = 2.
- AGE : Under 20 = 1; Twenties = 2; Thirties = 3; Forties = 4; Fifties = 5; Sixties = 6.
- EDUC: Grade = 1; Some High = 2; High = 3; Some College = 4; B.A. or B.S. = 5; M.A. or M.S. = 6; Ph.D. = 7.
- MSTAT: Single = 1; Married = 2.
- YEARS: 1 or less = 1; 1 to 4 = 2; 4 to 8 = 3; 8 to 12 = 4; 12 to 16 = 5; Over 16 = 6.

Table III.9: Firm 48 Demographics

VARIABLE	NATIONAL DATA		FIRM 48 DATA		STAT TEST RESULTS
	MEAN (M1)	STD.DEV. (V1)	MEAN (M2)	STD.DEV. (V2)	
SKILL VARIETY	5.408	1.106	5.508	1.285	
TASK IDENTITY	5.205	1.110	5.079	1.345	
TASK SIGNIFICANCE	5.605	1.176	5.476	1.289	
AUTONOMY	5.290	1.098	5.302	1.043	
FEEDBACK FROM JOB	5.133	1.066	5.512	1.091	
FDBK FROM SUPERVISORS	3.966	1.559	4.302	1.650	
GROWTH NEED STRENGTH	5.905	.994	6.214	.773	M2 GT M1 (.05)
MPS	153.581	62.464	167.255	79.607	V2 GT V1 (.05)

VARIABLE	ALL RESPONDENTS		FIRM 48 DATA		STAT TEST RESULTS
	MEAN (M1)	STD.DEV. (V1)	MEAN (M2)	STD.DEV. (V2)	
SKILL VARIETY	5.525	.983	5.508	1.285	V2 GT V1 (.05)
TASK IDENTITY	5.102	1.150	5.079	1.345	
TASK SIGNIFICANCE	5.486	1.146	5.476	1.289	
AUTONOMY	5.189	1.072	5.302	1.043	
FEEDBACK FROM JOB	5.005	1.104	5.512	1.091	M2 GT M1 (.05)
FDBK FROM SUPERVISORS	4.184	1.549	4.302	1.650	
GROWTH NEED STRENGTH	6.217	.771	6.214	.773	
MPS	146.609	64.292	167.255	79.607	

All variables except MPS are scored on a 1 to 7 scale.

Table III.10: Firm 48 Job Variables

VARIABLE	NATIONAL DATA		FIRM 50 DATA		STAT TEST RESULTS
	MEAN (M1)	STD.DEV. (V1)	MEAN (M2)	STD.DEV. (V2)	
SEX	1.262	.450	1.328	.473	
AGE	2.939	.913	3.284	.966	M2 GT M1 (.01)
EDUCATION	4.594	.702	4.045	.661	M2 LT M1 (.01)
MARITAL STATUS	1.626	.487	1.731	.479	M2 GT M1 (.05)
YEARS WITH FIRM	2.728	1.419	4.239	1.508	M2 GT M1 (.01)

VARIABLE	ALL RESPOND.		FIRM 50 DATA		STAT TEST RESULTS
	MEAN (M1)	STD.DEV. (V1)	MEAN (M2)	STD.DEV. (V2)	
SEX	1.342	.479	1.328	.473	
AGE	2.813	.945	3.284	.966	M2 GT M1 (.01)
EDUCATION	4.632	1.134	4.045	.661	M2 LT M1 (.01) V2 LT V1 (.01)
MARITAL STATUS	1.591	.544	1.731	.479	M2 GT M1 (.05)
YEARS WITH FIRM	2.804	1.695	4.239	1.508	M2 GT M1 (.01)

For National Data, N = 709
For All Respondents, N = 555
For This Firm, N = 67

NOTE...Significance levels are in parentheses in Results Column.

Variables are scored as follows:
SEX : Male = 1; Female = 2.
AGE : Under 20 = 1; Twenties = 2; Thirties = 3; Forties = 4; Fifties = 5; Sixties = 6.
EDUC: Grade = 1; Some High = 2; High = 3; Some College = 4; B.A. or B.S. = 5; M.A. or M.S. = 6; Ph.D. = 7.
MSTAT: Single = 1; Married = 2.
YEARS: 1 or less = 1; 1 to 4 = 2; 4 to 8 = 3; 8 to 12 = 4; 12 to 16 = 5; Over 16 = 6.

Table III.11: Firm 50 Demographics

VARIABLE	NATIONAL DATA		FIRM 50 DATA		STAT TEST RESULTS
	MEAN (M1)	STD.DEV. (V1)	MEAN (M2)	STD.DEV. (V2)	
SKILL VARIETY	5.408	1.106	5.453	1.005	
TASK IDENTITY	5.205	1.110	4.970	1.108	
TASK SIGNIFICANCE	5.605	1.176	5.537	1.225	
AUTONOMY	5.290	1.098	5.139	1.084	
FEEDBACK FROM JOB	5.133	1.066	5.030	1.154	
FDBK FROM SUPERVISORS	3.966	1.559	3.920	1.589	
GROWTH NEED STRENGTH	5.905	.994	5.945	1.107	
MPS	153.581	62.464	147.746	73.020	V2 LT V1 (.05)

VARIABLE	ALL RESPONDENTS		FIRM 50 DATA		STAT TEST RESULTS
	MEAN (M1)	STD.DEV. (V1)	MEAN (M2)	STD.DEV. (V2)	
SKILL VARIETY	5.525	.983	5.453	1.005	
TASK IDENTITY	5.102	1.150	4.970	1.108	
TASK SIGNIFICANCE	5.486	1.146	5.537	1.225	
AUTONOMY	5.189	1.072	5.139	1.084	
FEEDBACK FROM JOB	5.005	1.104	5.030	1.154	
FDBK FROM SUPERVISORS	4.184	1.549	3.920	1.589	
GROWTH NEED STRENGTH	6.217	.771	5.945	1.107	M2 LT M1 (.05) V2 GT V1 (.01)
MPS	146.609	64.292	147.746	73.020	V2 GT V1 (.05)

All variables except MPS are scored on a 1 to 7 scale.

Table III.12: Firm 50 Job Variables

VARIABLE	NATIONAL DATA		FIRM 491 DATA		STAT TEST RESULTS
	MEAN (M1)	STD.DEV. (V1)	MEAN (M2)	STD.DEV. (V2)	
SEX	1.262	.450	1.298	.459	
AGE	2.939	.913	2.725	.920	M2 LT M1 (.01)
EDUCATION	4.594	.702	4.931	1.076	M2 GT M1 (.01) V2 GT V1 (.01)
MARITAL STATUS	1.626	.487	1.595	.537	M2 LT M1 (.01)
YEARS WITH FIRM	2.728	1.419	2.595	1.654	

VARIABLE	ALL RESPOND.		FIRM 491 DATA		STAT TEST RESULTS
	MEAN (M1)	STD.DEV. (V1)	MEAN (M2)	STD.DEV. (V2)	
SEX	1.342	.479	1.298	.459	
AGE	2.813	.945	2.725	.920	
EDUCATION	4.632	1.134	4.931	1.076	M2 GT M1 (.01)
MARITAL STATUS	1.591	.544	1.595	.537	
YEARS WITH FIRM	2.804	1.695	2.595	1.654	

For National Data, N = 709
For All Respondents, N = 555
For This Firm, N = 131

NOTE...Significance levels are in parentheses in Results Column.

Variables are scored as follows:

SEX : Male = 1; Female = 2.
AGE : Under 20 = 1; Twenties = 2; Thirties = 3; Forties = 4; Fifties = 5; Sixties = 6.
EDUC: Grade = 1; Some High = 2; High = 3; Some College = 4; B.A. or B.S. = 5; M.A. or M.S. = 6; Ph.D. = 7.
MSTAT: Single = 1; Married = 2.
YEARS: 1 or less = 1; 1 to 4 = 2; 4 to 8 = 3; 8 to 12 = 4; 12 to 16 = 5; Over 16 = 6.

Table III.13: Firm 491 Demographics

VARIABLE	NATIONAL DATA		FIRM 491 DATA		STAT TEST RESULTS
	MEAN (M1)	STD.DEV. (V1)	MEAN (M2)	STD.DEV. (V2)	
SKILL VARIETY	5.408	1.106	5.567	.930	M2 GT M1 (.05) V2 LT V1 (.01)
TASK IDENTITY	5.205	1.110	5.354	.979	V2 LT V1 (.05)
TASK SIGNIFICANCE	5.605	1.176	5.476	1.096	
AUTONOMY	5.290	1.098	5.219	.962	V2 LT V1 (.05)
FEEDBACK FROM JOB	5.133	1.066	5.011	1.058	
FDBK FROM SUPERVISORS	3.966	1.559	4.438	1.417	M2 GT M1 (.01)
GROWTH NEED STRENGTH	5.905	.994	6.243	.740	M2 GT M1 (.01) V2 LT V1 (.01)
MPS	153.581	62.464	147.343	57.816	

VARIABLE	ALL RESPONDENTS		FIRM 491 DATA		STAT TEST RESULTS
	MEAN (M1)	STD.DEV. (V1)	MEAN (M2)	STD.DEV. (V2)	
SKILL VARIETY	5.525	.983	5.567	.930	
TASK IDENTITY	5.102	1.150	5.354	.979	M2 GT M1 (.01) V2 LT V1 (.01)
TASK SIGNIFICANCE	5.486	1.146	5.476	1.096	
AUTONOMY	5.189	1.072	5.219	.962	V2 LT V1 (.05)
FEEDBACK FROM JOB	5.005	1.104	5.011	1.058	
FDBK FROM SUPERVISORS	4.184	1.549	4.438	1.417	M2 GT M1 (.05)
GROWTH NEED STRENGTH	6.217	.771	6.243	.740	
MPS	146.609	64.292	147.343	57.816	V2 LT V1 (.05)

All variables except MPS are scored on a 1 to 7 scale.

Table III.14: Firm 491 Job Variables

VARIABLE	NATIONAL DATA		FIRM 492 DATA		STAT TEST RESULTS
	MEAN (M1)	STD.DEV. (V1)	MEAN (M2)	STD.DEV. (V2)	
SEX	1.262	.450	1.278	.449	
AGE	2.939	.913	2.883	1.018	V2 GT V1 (.05)
EDUCATION	4.594	.702	4.611	1.424	V2 GT V1 (.01)
MARITAL STATUS	1.626	.487	1.568	.545	V2 GT V1 (.05)
YEARS WITH FIRM	2.728	1.419	3.031	1.836	M2 GT M1 (.05) V2 GT V1 (.01)

VARIABLE	ALL RESPOND.		FIRM 492 DATA		STAT TEST RESULTS
	MEAN (M1)	STD.DEV. (V1)	MEAN (M2)	STD.DEV. (V2)	
SEX	1.342	.479	1.278	.449	
AGE	2.813	.945	2.883	1.018	
EDUCATION	4.632	1.134	4.611	1.424	V2 GT V1 (.01)
MARITAL STATUS	1.591	.544	1.568	.545	
YEARS WITH FIRM	2.804	1.695	3.031	1.836	

For National Data, N = 709
For All Respondents, N = 555
For This Firm, N = 162

NOTE...Significance levels are in parentheses in Results Column.

Variables are scored as follows:

SEX : Male = 1; Female = 2.
AGE : Under 20 = 1; Twenties = 2; Thirties = 3; Forties = 4; Fifties = 5; Sixties = 6.
EDUC: Grade = 1; Some High = 2; High = 3; Some College = 4; B.A. or B.S. = 5; M.A. or M.S. = 6; Ph.D. = 7.
MSTAT: Single = 1; Married = 2.
YEARS: 1 or less = 1; 1 to 4 = 2; 4 to 8 = 3; 8 to 12 = 4; 12 to 16 = 5; Over 16 = 6.

Table III.15: Firm 492 Demographics

VARIABLE	NATIONAL DATA		FIRM 492 DATA		STAT TEST RESULTS
	MEAN (M1)	STD.DEV. (V1)	MEAN (M2)	STD.DEV. (V2)	
SKILL VARIETY	5.408	1.106	5.689	.907	M2 GT M1 (.01) V2 LT V1 (.01)
TASK IDENTITY	5.205	1.110	5.241	1.104	
TASK SIGNIFICANCE	5.605	1.176	5.539	1.043	V2 LT V1 (.05)
AUTONOMY	5.290	1.098	5.342	1.021	
FEEDBACK FROM JOB	5.133	1.066	5.136	.998	
FDBK FROM SUPERVISORS	3.966	1.559	4.171	1.557	
GROWTH NEED STRENGTH	5.905	.994	6.302	.695	M2 GT M1 (.01) V2 LT V1 (.01)
MPS	153.581	62.464	156.629	61.645	

VARIABLE	ALL RESPONDENTS		FIRM 492 DATA		STAT TEST RESULTS
	MEAN (M1)	STD.DEV. (V1)	MEAN (M2)	STD.DEV. (V2)	
SKILL VARIETY	5.525	.983	5.689	.907	M2 GT M1 (.05)
TASK IDENTITY	5.102	1.150	5.241	1.104	
TASK SIGNIFICANCE	5.486	1.146	5.539	1.043	
AUTONOMY	5.189	1.072	5.342	1.021	M2 GT M1 (.05)
FEEDBACK FROM JOB	5.005	1.104	5.136	.998	V2 LT V1 (.05)
FDBK FROM SUPERVISORS	4.184	1.549	4.171	1.557	
GROWTH NEED STRENGTH	6.217	.771	6.302	.695	V2 LT V1 (.05)
MPS	146.609	64.292	156.629	61.645	M2 GT M1 (.05)

All variables except MPS are scored on a 1 to 7 scale.

Table III.16: Firm 492 Job Variables

APPENDIX IV:
DETAILED MAINTENANCE RESPONSES

VARIABLE	NATIONAL DATA.		ALL RESPONDENTS		STAT TEST RESULTS
	MEAN (M1)	STD.DEV. (V1)	MEAN (M2)	STD.DEV. (V2)	
SEX	1.262	.450	1.304	.469	
AGE	2.939	.913	3.019	.968	
EDUCATION	4.594	.702	4.700	1.160	V2 GT V1 (.01)
MARITAL STATUS	1.626	.487	1.650	.524	
YEARS WITH FIRM	2.728	1.419	3.100	1.739	M2 GT M1 (.01) V2 GT V1 (.01)

For National Data, N = 709
For Maintenance Data, N = 260

NOTE...Significance levels are in parentheses in Results Column.

Variables are scored as follows:

SEX : Male = 1; Female = 2.
AGE : Under 20 = 1; Twenties = 2; Thirties = 3; Forties = 4; Fifties = 5; Sixties = 6.
EDUC: Grade = 1; Some High = 2; High = 3; Some College = 4; B.A. or B.S. = 5; M.A. or M.S. = 6; Ph.D. = 7.
MSTAT: Single = 1; Married = 2.
YEARS: 1 or less = 1; 1 to 4 = 2; 4 to 8 = 3; 8 to 12 = 4; 12 to 16 = 5; Over 16 = 6.

PROGRAMMER AND ANALYST DEMOGRAPHICS
National Data vs. Those With (Maint. LE 20%)
Table IV.1

VARIABLE	NATIONAL DATA.		ALL RESPONDENTS		STAT TEST RESULTS
	MEAN (M1)	STD.DEV. (V1)	MEAN (M2)	STD.DEV. (V2)	
SEX	1.262	.450	1.369	.485	M2 GT M1 (.05)
AGE	2.939	.913	2.650	.871	M2 LT M1 (.01)
EDUCATION	4.594	.702	4.495	1.187	V2 GT V1 (.01)
MARITAL STATUS	1.626	.487	1.583	.496	
YEARS WITH FIRM	2.728	1.419	2.835	1.541	

For National Data, N = 709
For Maintenance Data, N = 103

NOTE...Significance levels are in parentheses in Results Column.

Variables are scored as follows:
SEX : Male = 1; Female = 2.
AGE : Under 20 = 1; Twenties = 2; Thirties = 3; Forties = 4; Fifties = 5; Sixties = 6.
EDUC: Grade = 1; Some High = 2; High = 3; Some College = 4; B.A. or B.S. = 5; M.A. or M.S. = 6; Ph.D. = 7.
MSTAT: Single = 1; Married = 2.
YEARS: 1 or less = 1; 1 to 4 = 2; 4 to 8 = 3; 8 to 12 = 4; 12 to 16 = 5; Over 16 = 6.

PROGRAMMER AND ANALYST DEMOGRAPHICS
National Data vs. Those With (20% LT Maint. LE 40%)
Table IV.2

VARIABLE	NATIONAL DATA.		ALL RESPONDENTS		STAT TEST RESULTS
	MEAN (M1)	STD.DEV. (V1)	MEAN (M2)	STD.DEV. (V2)	
SEX	1.262	.450	1.396	.492	M2 GT M1 (.01)
AGE	2.939	.913	2.772	.968	
EDUCATION	4.594	.702	4.545	1.171	V2 GT V1 (.01)
MARITAL STATUS	1.626	.487	1.594	.586	V2 GT V1 (.05)
YEARS WITH FIRM	2.728	1.419	2.614	1.755	V2 GT V1 (.01)

For National Data, N = 709
For Maintenance Data, N = 101

NOTE...Significance levels are in parentheses in Results Column.

Variables are scored as follows:
SEX : Male = 1; Female = 2.
AGE : Under 20 = 1; Twenties = 2; Thirties = 3; Forties = 4; Fifties = 5; Sixties = 6.
EDUC: Grade = 1; Some High = 2; High = 3; Some College = 4; B.A. or B.S. = 5; M.A. or M.S. = 6; Ph.D. = 7.
MSTAT: Single = 1; Married = 2.
YEARS: 1 or less = 1; 1 to 4 = 2; 4 to 8 = 3; 8 to 12 = 4; 12 to 16 = 5; Over 16 = 6.

PROGRAMMER AND ANALYST DEMOGRAPHICS
National Data vs. Those With (40% LT Maint. LE 60%)
Table IV.3

VARIABLE	NATIONAL DATA.		ALL RESPONDENTS		STAT TEST RESULTS
	MEAN (M1)	STD.DEV. (V1)	MEAN (M2)	STD.DEV. (V2)	
SEX	1.262	.450	1.327	.474	
AGE	2.939	.913	2.449	.867	M2 LT M1 (.01)
EDUCATION	4.594	.702	4.653	1.200	V2 GT V1 (.01)
MARITAL STATUS	1.626	.487	1.469	.544	M2 LT M1 (.05)
YEARS WITH FIRM	2.728	1.419	2.306	1.661	M2 LT M1 (.05)

For National Data, N = 709
For Maintenance Data, N = 49

NOTE...Significance levels are in parentheses in Results Column.

Variables are scored as follows:

SEX : Male = 1; Female = 2.
AGE : Under 20 = 1; Twenties = 2; Thirties = 3; Forties = 4; Fifties = 5; Sixties = 6.
EDUC: Grade = 1; Some High = 2; High = 3; Some College = 4; B.A. or B.S. = 5; M.A. or M.S. = 6; Ph.D. = 7.
MSTAT: Single = 1; Married = 2.
YEARS: 1 or less = 1; 1 to 4 = 2; 4 to 8 = 3; 8 to 12 = 4; 12 to 16 = 5; Over 16 = 6.

PROGRAMMER AND ANALYST DEMOGRAPHICS
National Data vs. Those With (60% LT Maint. LE 80%)
Table IV.4

VARIABLE	NATIONAL DATA.		ALL RESPONDENTS		STAT TEST RESULTS
	MEAN (M1)	STD.DEV. (V1)	MEAN (M2)	STD.DEV. (V2)	
SEX	1.262	.450	1.405	.497	M2 GT M1 (.05)
AGE	2.939	.913	2.452	.670	M2 LT M1 (.01) V2 LT V1 (.01)
EDUCATION	4.594	.702	4.738	.497	M2 GT M1 (.05) V2 LT V1 (.01)
MARITAL STATUS	1.626	.487	1.381	.623	M2 LT M1 (.01) V2 GT V1 (.01)
YEARS WITH FIRM	2.728	1.419	1.929	1.156	M2 LT M1 (.01) V2 LT V1 (.05)

For National Data, N = 709
For Maintenance Data, N = 42

NOTE...Significance levels are in parentheses in Results Column.

Variables are scored as follows:

SEX : Male = 1; Female = 2.
AGE : Under 20 = 1; Twenties = 2; Thirties = 3; Forties = 4; Fifties = 5; Sixties = 6.
EDUC: Grade = 1; Some High = 2; High = 3; Some College = 4; B.A. or B.S. = 5; M.A. or M.S. = 6; Ph.D. = 7.
MSTAT: Single = 1; Married = 2.
YEARS: 1 or less = 1; 1 to 4 = 2; 4 to 8 = 3; 8 to 12 = 4; 12 to 16 = 5; Over 16 = 6.

PROGRAMMER AND ANALYST DEMOGRAPHICS
National Data vs. Those With (Maint. GT 80%)
Table IV.5

VARIABLE	NATIONAL DATA.		ALL RESPONDENTS		STAT TEST RESULTS
	MEAN (M1)	STD.DEV. (V1)	MEAN (M2)	STD.DEV. (V2)	
SEX	1.262	.450	1.378	.486	M2 GT M1 (.01)
AGE	2.939	.913	2.622	.901	M2 LT M1 (.01)
EDUCATION	4.594	.702	4.595	1.080	V2 GT V1 (.01)
MARITAL STATUS	1.626	.487	1.519	.591	M2 LT M1 (.01) V2 GT V1 (.01)
YEARS WITH FIRM	2.728	1.419	2.395	1.645	M2 LT M1 (.05) V2 GT V1 (.01)

For National Data, N = 709
For Maintenance Data, N = 185

NOTE...Significance levels are in parentheses in Results Column.

Variables are scored as follows:
SEX : Male = 1; Female = 2.
AGE : Under 20 = 1; Twenties = 2; Thirties = 3; Forties = 4; Fifties = 5; Sixties = 6.
EDUC: Grade = 1; Some High = 2; High = 3; Some College = 4; B.A. or B.S. = 5; M.A. or M.S. = 6; Ph.D. = 7.
MSTAT: Single = 1; Married = 2.
YEARS: 1 or less = 1; 1 to 4 = 2; 4 to 8 = 3; 8 to 12 = 4; 12 to 16 = 5; Over 16 = 6.

PROGRAMMER AND ANALYST DEMOGRAPHICS
National Data vs. Those With (GE 50% Maint.)
Table IV.6

VARIABLE	NATIONAL DATA		ALL RESPONDENTS		STAT TEST RESULTS
	MEAN (M1)	STD.DEV. (V1)	MEAN (M2)	STD.DEV. (V2)	
SKILL VARIETY	5.408	1.106	5.678	.995	M2 GT M1 (.01)
TASK IDENTITY	5.205	1.110	5.306	1.161	
TASK SIGNIFICANCE	5.605	1.176	5.537	1.177	
AUTONOMY	5.290	1.098	5.382	1.095	
FEEDBACK FROM JOB	5.133	1.066	5.116	1.058	
FDBK FROM SUPERVISORS	3.966	1.559	4.255	1.609	M2 GT M1 (.01)
GROWTH NEED STRENGTH	5.905	.994	6.296	.681	M2 GT M1 (.01) V2 LT V1 (.01)
MPS	153.581	62.464	158.985	67.265	

For National Data, N = 709
For Maintenance Data, N = 260

NOTE...Significance levels are in parentheses in Results Column.

All variables except MPS are scored on a 1 to 7 scale.

PROGRAMMER AND ANALYST JOB VARIABLES
National Data vs. Those With (Maint. LE 20%)
Table IV.7

VARIABLE	NATIONAL DATA		ALL RESPONDENTS		STAT TEST RESULTS
	MEAN (M1)	STD.DEV. (V1)	MEAN (M2)	STD.DEV. (V2)	
SKILL VARIETY	5.408	1.106	5.586	.782	M2 GT M1 (.05) V2 LT V1 (.01)
TASK IDENTITY	5.205	1.110	5.204	1.091	
TASK SIGNIFICANCE	5.605	1.176	5.450	1.142	
AUTONOMY	5.290	1.098	5.149	1.083	
FEEDBACK FROM JOB	5.133	1.066	4.990	1.095	
FDBK FROM SUPERVISORS	3.966	1.559	4.288	1.484	
GROWTH NEED STRENGTH	5.905	.994	6.146	.710	M2 GT M1 (.01) V2 LT V1 (.01)
MPS	153.581	62.464	145.292	61.417	

For National Data, N = 709
For Maintenance Data, N = 103

NOTE...Significance levels are in parentheses in Results Column.

All variables except MPS are scored on a 1 to 7 scale.

PROGRAMMER AND ANALYST JOB VARIABLES
National Data vs. Those With (20% LT Maint. LE 40%)
Table IV.8

VARIABLE	NATIONAL DATA		ALL RESPONDENTS		STAT TEST RESULTS
	MEAN (M1)	STD.DEV. (V1)	MEAN (M2)	STD.DEV. (V2)	
SKILL VARIETY	5.408	1.106	5.406	.961	V2 LT V1 (.01)
TASK IDENTITY	5.205	1.110	4.785	1.002	M2 LT M1 (.01) V2 LT V1 (.05)
TASK SIGNIFICANCE	5.605	1.176	5.409	1.187	
AUTONOMY	5.290	1.098	5.030	.927	M2 LT M1 (.01) V2 LT V1 (.01)
FEEDBACK FROM JOB	5.133	1.066	5.000	1.117	
FDBK FROM SUPERVISORS	3.966	1.559	4.188	1.495	
GROWTH NEED STRENGTH	5.905	.994	6.069	.996	
MPS	153.581	62.464	136.291	57.284	M2 LT M1 (.01)

For National Data, N = 709
For Maintenance Data, N = 101

NOTE...Significance levels are in parentheses in Results Column.

All variables except MPS are scored on a 1 to 7 scale.

PROGRAMMER AND ANALYST JOB VARIABLES
National Data vs. Those With (40% LT Maint. LE 60%)
Table IV.9

VARIABLE	NATIONAL DATA		ALL RESPONDENTS		STAT TEST RESULTS
	MEAN (M1)	STD.DEV. (V1)	MEAN (M2)	STD.DEV. (V2)	
SKILL VARIETY	5.408	1.106	5.395	.935	
TASK IDENTITY	5.205	1.110	5.109	.951	
TASK SIGNIFICANCE	5.605	1.176	5.497	.900	V2 LT V1 (.01)
AUTONOMY	5.290	1.098	5.020	1.057	M2 LT M1 (.05)
FEEDBACK FROM JOB	5.133	1.066	4.990	1.021	
FDBK FROM SUPERVISORS	3.966	1.559	4.129	1.461	
GROWTH NEED STRENGTH	5.905	.994	6.167	.919	M2 GT M1 (.05)
MPS	153.581	62.464	138.994	55.886	M2 LT M1 (.05)

For National Data, N = 709
For Maintenance Data, N = 49

NOTE...Significance levels are in parentheses in Results Column.

All variables except MPS are scored on a 1 to 7 scale.

PROGRAMMER AND ANALYST JOB VARIABLES
National Data vs. Those With (60% LT Maint. LE 80%)
Table IV.10

VARIABLE	NATIONAL DATA		ALL RESPONDENTS		STAT TEST RESULTS
	MEAN (M1)	STD.DEV. (V1)	MEAN (M2)	STD.DEV. (V2)	
SKILL VARIETY	5.408	1.106	4.865	1.157	M2 LT M1 (.01)
TASK IDENTITY	5.205	1.110	4.341	1.324	M2 LT M1 (.01) V2 GT V1 (.05)
TASK SIGNIFICANCE	5.605	1.176	5.437	1.149	
AUTONOMY	5.290	1.098	4.675	1.011	M2 LT M1 (.01)
FEEDBACK FROM JOB	5.133	1.066	4.375	1.289	M2 LT M1 (.01) V2 GT V1 (.05)
FDBK FROM SUPERVISORS	3.966	1.559	3.540	1.469	M2 LT M1 (.05)
GROWTH NEED STRENGTH	5.905	.994	6.321	.683	M2 GT M1 (.01) V2 LT V1 (.01)
MPS	153.581	62.464	106.928	57.486	M2 LT M1 (.01)

For National Data, N = 709
For Maintenance Data, N = 42

NOTE...Significance levels are in parentheses in Results Column.

All variables except MPS are scored on a 1 to 7 scale.

PROGRAMMER AND ANALYST JOB VARIABLES
National Data vs. Those With (Maint. GT 80%)
Table IV.11

VARIABLE	NATIONAL DATA		ALL RESPONDENTS		STAT TEST RESULTS
	MEAN (M1)	STD.DEV. (V1)	MEAN (M2)	STD.DEV. (V2)	
SKILL VARIETY	5.408	1.106	5.256	1.021	M2 LT M1 (.05)
TASK IDENTITY	5.205	1.110	4.760	1.094	M2 LT M1 (.01)
TASK SIGNIFICANCE	5.605	1.176	5.405	1.111	M2 LT M1 (.05)
AUTONOMY	5.290	1.098	4.923	.987	M2 LT M1 (.01)
FEEDBACK FROM JOB	5.133	1.066	4.826	1.156	M2 LT M1 (.01)
FDBK FROM SUPERVISORS	3.966	1.559	4.014	1.493	
GROWTH NEED STRENGTH	5.905	.994	6.139	.910	M2 GT M1 (.01)
MPS	153.581	62.464	128.242	57.288	M2 LT M1 (.01)

For National Data, N = 709
For Maintenance Data, N = 185

NOTE...Significance levels are in parentheses in Results Column.

All variables except MPS are scored on a 1 to 7 scale.

PROGRAMMER AND ANALYST JOB VARIABLES
National Data vs. Those With (GE 50% Maint.)
Table IV.12

APPENDIX V: BIBLIOGRAPHY

Boehm, Barry W., Software Engineering Economics, 1981, Prentice-Hall, Inc., Englewood Cliffs, NJ, p. 18.

——"Software Engineering," IEEE Transactions on Computers, December, 1976, pp. 1226-1241.

Couger, J. Daniel, "Analysis of Key Factors for Motivation of Data Processing Professionals," Proceedings, 6th New Zealand Conference, Auckland, August, 1978, pp. 1-19.

——"Motivators/Demotivators for DP Professionals," Proceedings, CIPS/DPMA/FIQ, June, 1979, Quebec, Canada, pp. 292-298.

——"ACPA Members Rate Their Jobs High in Motivation," Computerworld, December 3, 1979, pp. 31-32.

——"What Motivates MIS Managers," Computerworld, March 10, 1980, pp. 9-16 (special In Depth center section).

——"Training and Education of Computer Professionals in the USA," Proceedings, Seminar on Computer Education in Singapore, Science Council of Singapore, October 2-3, 1980, pp. 1-118.

——"The Project Manager: Kingpin in Personnel Motivation," Computerworld Extra, September 1, 1981, pp. 49-55.

——"Improving the Productivity of DP Trainers," Computing Newsletter, February 1981, pp. 1-8.

Couger, J. Daniel and Robert A. Zawacki, Motivating and Managing Computer Personnel, Wiley-Interscience, 1980.

——"What Motivates DP Professionals," Datamation, September, 1978, pp. 116-123.

——"Compensation Preferences of DP Professionals," Datamation, November, 1978, pp. 94-102.

——"Something's Very Wrong With DP Operations Jobs," Datamation, March, 1979, pp. 149-158.

——"Studies Show Problems With DPers' Motivation," Computerworld, May 7, 1979, pp. 26, 30.

——"Motivation Levels for Directors of University Computer Centers vs. Those of Their Employees," EDUCOM Bulletin, Vol. 14, No. 4, Winter, 1979, pp. 2-8.

——"Motivation Levels of MIS Managers vs. Those of Their Employees," MIS Quarterly, September, 1979, pp. 47-56.

Daly, E. B., "Management of Software Development," IEEE Transactions on Software Engineering, May, 1977, pp. 229-242.

Diebold, J., "Improving the Utilization of Personnel Resources," The Diebold Computer Planning and Management Service, August, 1979, pp. 44-46.

Elshoff, J. L., "An Analysis of Some Commercial PL/1 Programs," IEEE Transactions on Software Engineering, June, 1976, pp. 113-120.

Flamboltz, E. G., "Toward a Theory of Human Resource Value Accounting in Formal Organizations," The Accounting Review, Vol. 47, 1972, pp. 666-678.

"Federal Agencies' Maintenance of Computer Programs: Expensive and Undermanaged," U.S. General Accounting Office, Gaithersburg, MD, 1981.

Fitz-enz, Jac, "Who is the DP Professional?" Datamation, September, 1978, pp. 125-128.

Glass, Robert L. and Ronald A. Noiseux, Software Maintenance Guidebook, Prentice-Hall, Inc., Englewood Cliffs, NJ, 1981.

Glover, John Desmond, and Ralph M. Hower, "Observations on the Lincoln Electric Company," The Administrator, Richard D. Irwin, Inc., Homewood, IL, 1963, pp. 243-288.

Hackman, J. R. and E. E. Lawler, "Employee Reactions to Job Characteristics," Journal of Applied Psychology Monograph, 1971, pp. 259-286.

Hackman, J. R., G. R. Oldham, R. Janson, and K. Purdy, "A New Strategy for Job Enrichment," California Management Review, Vol. 17, No. 4, 1975, p. 58-60.

Hackman, J. R. and G. R. Oldham, "Development of the Job Diagnostic Survey," Journal of Applied Psychology, Vol. 60, No. 2, 1975, pp. 159-170.

Herzberg, Frederick, The Motivation to Work, John Wiley and Sons, Inc., New York, 1959.

Howard, Philip, "Examining the Maintenance Issue," System Development, April, 1982, p. 4.

Lientz, Bennet P. and Burton E. Swanson, Software Maintenance Management, Addison-Wesley Publishing Co., Inc., Reading, MA, 1980.

Lyons, M. J., "Structured Retrofit—1980," 1980 Proceedings of SHARE 55, 1980, pp. 263-265

Martin, James and Carma L. McClure, Software Maintenance: The Problem and Its Solutions, Prentice-Hall, Inc., Englewood Cliffs, NJ, 1983.

Maslow, Abraham, Motivation and Personality, New York: Harper & Row, 1954.

McClure, Carma L., Managing Software Development and Maintenance, Van Nostrand Reinhold Co., New York, NY, 1981.

McLaughlin, R. A., "That Old Bugaboo, Turnover," Datamation, October, 1979, p. 96.

Nie, N. H. et al., Statistical Package for the Social Sciences, McGraw-Hill, 1975, pp. 269-270.

Parikh, Girish, ed., Techniques of Program and System Maintenance, Little, Brown and Company, Boston, MA, 1982.

Parikh, Girish and Nicholas Zvegintzov, eds., Tutorial on Software Maintenance, IEEE Computer Society, Long Beach, CA, 1983.

Perry, William E., Managing Systems Maintenance, Q.E.D. Information Sciences, Inc., Wellesley, MA, 1981.

Swanson, E. "The Dimensions of Maintenance," 2nd International Conference In Software Engineering, Oct. 1976.

Turner, A. N., and P. R. Lawrence, Industrial Jobs and the Worker, Harvard Graduate School of Business Administration, 1965.

"What Do Programmers Want?" Output, February, 1981, pp. 56-61.

ARTICLES WHERE OUR WORK IS DISCUSSED

Bruce Howard, "Trainers' Desire for Promotion Held Greater Than Students'," Computerworld, January 12, 1981, p. 20.

Charles S. Clifton, "Computer Techs are People, Too," Colorado Springs Sun, January 14, 1981, p. 3F.

"Whither DP Trainers," Northeast Training News, January, 1981, pp. 6-7.

Bruce Howard, "Productivity Woes Tied to Societal Explosions," Computerworld, February 16, 1981, p. 12.

Ann Dooley, "Turnover: Myth of Impediment to Managers?" Computerworld, February 25, 1981, p. 4.

Robert Batt, "Putting on a New Face: The Need for Behavioral Skills in MIS," Computerworld, December 28/January 4, 1982, pp. 77-81.

Josh Martin, "Choosing a Management Style," Computer Decisions, December, 1981, pp. 81-86, 146-153.

A. L. LeDuc, Jr., "Motivation of Programmers," Database, Summer, 1980, pp. 4-12.

Martin Lasden, "Cut Turnover With a Japanese Pattern," Computer Decisions, September, 1981, pp. 135-148.

Phillip C. Howard, "Motivation of Personnel," System Development, July, 1981, pp. 3-4.

Richard D. Canning, "The Challenge of Increased Productivity," EDP Analyzer, April, 1981, pp. 1-12.

Bruce Howard, "Study: It's Not Money That Motivates DPers," Computerworld, May 26, 1980.

Jeffry Beeler, "Seminar: Meaningful Work, Job Interest Top Priorities in Boosting DP Productivity," Computerworld, April 7, 1980, p. 9.

Jeffry Beeler, "Chitchat Seen Vital for MIS Heads," Computerworld, April 7, 1980, p. 2.

"Study of Jobs and the People Who Hold Them," Sitebytes, Autumn, 1979, p. 9.

"ACPA Members Rate Jobs High in Motivation," Computerworld, December 3, 1979, pp. 31-32.

Nancy French, "Team Approach May Cut Programmer Output," Computerworld, October 9, 1979, p. 7.

Jeffry Beeler, "Operations Staff's Career Needs Seen Neglected," Computerworld, April 14, 1980, p. 9.

INDEX

J

K

L

M

N

U

V

W